UNLOCKING SCIENCE PROCESS SKILLS

GLOBE FEARON
Pearson Learning Group

Art and Design: Evelyn Bauer, Tricia Battipede, Tracey Gerber, Jenifer Hixson, Judy Mahoney, Dave Mager, Elbaliz Mendez
Editorial: Stephanie P. Cahill, Maurice Sabean, Theresa McCarthy
Manufacturing: Mark Cirillo, Tom Dunne
Marketing: Douglas Falk, Maureen Christensen
Production: Irene Belinsky, Karen Edmonds, Leslie Greenberg, Alia Lesser, Cindy Talocci
Publishing Operations: Carolyn Coyle, Thomas Daning, Richetta Lobban

Consultants
Elizabeth Jimenez
Pamona, CA

Sean Devine
Science Teacher
Waldwick High School
Waldwick, New Jersey

Content Reviewer
Todd Woerner
Department of Chemistry
Duke University
Durham, NC

Teacher Reviewers
Peggy L. Cook
Lakeworth Middle School
Lakeworth, FL

Karen DeLaCroix
Science Department Chairperson
Garfield High School
Akron, OH

Trannie Sehlmeyer
Science Teacher
Start High School
Toledo, OH

About the cover: The black-collared lizard lives in a desert environment such as the one found in Monument Valley, Arizona (background). One might ask if the dry, hot climate affects the lizard's growth. As a scientist, you would observe, question, hypothesize, collect data, conclude, and report. You would probably use tools such as the caliper and barometer pictured to collect data. A graph in your report would help to clearly illustrate some of the information you found.

Photo Credits
All photography by Pearson Learning Group unless otherwise noted.
Front cover: *bkgd.* Bill Brooks/Wonderfile; *l.* © E.R. Degginger/Color-Pic, Inc.; *r.* David Toase/PhotoDisc/Getty Images, Inc.; *b.* Steven Cole/PhotoDisc/Getty Images, Inc. **Back cover:** Siede Preis/Getty Images, Inc.
Front and Back Matter Banner: *l.* © Corbis; *m.l.* Tony Freeman/Index Stock Imagery, Inc.; *m.r.* Charles Thatcher/Getty Images, Inc.; *r.* AP/Wide World Photo. 2: © Corbis. 2–3: Princeton University Plasma Physics Laboratory. 3: © Corbis. 4: *t.* © Corbis; *b.* © Kevin Fleming/Corbis. 5: *l.* © Patti Murray/Animals Animals/Earth Scenes; *r.* © R. Bauer/Photo Researchers, Inc. 7: E.R. Degginger/Color-Pic, Inc. 8: © Corbis. 12: © Corbis. 16: *t.* © Corbis; *b.* © E.R. Degginger/Color-Pic, Inc. 17: *t.* Official U.S. Navy Photo. 20: © Corbis. 24: © Corbis. 26: *l.* Jess Alford/Getty Images/PhotoDisc, Inc.; *m.* Getty Images, Inc.; *r.* Mark Segal/Index Stock Imagery, Inc. 28: *t.* © Corbis; *b.* © Mark A. Johnson/Stock Market/Corbis. 31: © E.R. Degginger/Color-Pic, Inc. 32: © Corbis. 36: *t.* © Corbis; *b.* © E.R. Degginger/Color-Pic, Inc. 38: Tony Freeman/Index Stock Imagery, Inc. 38–39: G. Brad Lewis/Stone/Getty Images, Inc. 39: Tony Freeman/Index Stock Imagery, Inc. 40: Tony Freeman/Index Stock Imagery, Inc. 44: Tony Freeman/Index Stock Imagery, Inc. 48: Tony Freeman/Index Stock Imagery, Inc. 52: Tony Freeman/Index Stock Imagery, Inc. 56: Tony Freeman/Index Stock Imagery, Inc. 59: *l.* Fernand Ivaldi/Getty Images, Inc.; *r.* AP Photo/NASA, Jet Propulsion Laboratories. 60: *t.* Tony Freeman/Index Stock Imagery, Inc.; *b.* © Richard R. Hansen/Photo Researchers, Inc.; 64: Tony Freeman/Index Stock Imagery, Inc. 68: Tony Freeman/Index Stock Imagery, Inc. 70: Charles Thatcher/Getty Images, Inc. 70–71: Michael Nichols/National Geographic Society. 71: Charles Thatcher/Getty Images, Inc. 72: *t.* Charles Thatcher/Getty Images, Inc.; *l.* Gerben Oppermans/Getty Images, Inc.; *m.* © Phil Degginger/Color-Pic, Inc.; *r.* © Phil Degginger/Color-Pic, Inc. 73: John Mead/Science Photo Library/Photo Researchers, Inc. 74: Peter Gregg/Color-Pic, Inc. 76: Charles Thatcher/Getty Images, Inc. 80: Charles Thatcher/Getty Images, Inc. 84: Charles Thatcher/Getty Images, Inc. 86: *l.* © E.R. Degginger/Color-Pic, Inc.; *m.l.* © Ken Cole/Animals Animals/Earth Scenes; *m.r.* © E.R. Degginger/Color-Pic, Inc.; *r.* © E.R. Degginger/Color-Pic, Inc. 88: Charles Thatcher/Getty Images, Inc. 92: *t.* Charles Thatcher/Getty Images, Inc.; *l.* Getty Images, Inc.; *r.* National Oceanic and Atmospheric Administration. 94: © Wesley Bocxe/Photo Researchers, Inc. 96: Charles Thatcher/Getty Images, Inc. 100: Charles Thatcher/Getty Images, Inc. 106: AP/Wide World Photo. 106–107: © Roger Ressmeyer/Corbis. 107: AP/Wide World Photo. 108: AP/Wide World Photo. 112: AP/Wide World Photo. 116: AP/Wide World Photo. 122: AP/Wide World Photo. 126: AP/Wide World Photo. 130: AP/Wide World Photo. 132: *t.l.* © Gary Conner/PhotoEdit; *t.r.* © David Young-Wolff/PhotoEdit; *b.* © Joyce Photographics/Photo Researchers, Inc. 133: *t.l.* © Michael Rosenfeld/Getty Images, Inc.; *t.r.* Elena Rooraid, PhotoEdit; *m.l.* © M. Meadows/Photo Researchers, Inc.; *m.r.* John Hill/Getty Images, Inc.; *b.l.* Keith Brofsky/Getty Images, Inc./PhotoDisc, Inc.; *b.r.* Andy Manis/AP/Wide World Photo.

Copyright © 2003 by Pearson Education, Inc., publishing as Globe Fearon®, an imprint of Pearson Learning Group, 299 Jefferson Road, Parsippany, NJ 07054. All rights reserved. No part of this book may be reproduced or transmitted in any form or by any means, electronic or mechanical, including photocopying, recording, or by any information storage and retrieval system, without permission in writing from the publisher. For information regarding permission(s), write to Rights and Permissions Department.

Globe Fearon® is a registered trademark of Globe Fearon, Inc.

ISBN 0-13-024081-8

Printed in the United States of America

2 3 4 5 6 7 8 9 10   10 09 08 07 06 05 04 03

1-800-321-3106
www.pearsonlearning.com

CONTENTS

To the Student 1

Chapter 1 ✦ Planning As a Scientist 2
Lesson 1 Making Observations and Interpretations 4
Lesson 2 Using Information Sources 8
Lesson 3 Organizing Information 12
Lesson 4 Reading in Science 16
Lesson 5 Asking Questions in Science 20
Lesson 6 Making Predictions 24
Lesson 7 Forming Hypotheses for Testing 28
Lesson 8 Planning an Experiment 32
Chapter 1 REVIEW 36

Chapter 2 ✦ Working As a Scientist 38
Lesson 1 Measuring 40
Lesson 2 Estimating Measurements 44
Lesson 3 Using the Metric System 48
Lesson 4 Practicing Safety in the Laboratory 52
Lesson 5 Working in the Laboratory 56
Lesson 6 Sampling 60
Lesson 7 Recording 64
Chapter 2 REVIEW 68

Chapter 3 ✦ Thinking As a Scientist 70
Lesson 1 Comparing 72
Lesson 2 Classifying 76
Lesson 3 Using Chemical Shorthand 80
Lesson 4 Using Guides and Keys 84
Lesson 5 Recognizing Patterns in Science 88
Lesson 6 Understanding Cause and Effect 92
Lesson 7 Concluding 96
Lesson 8 Generalizing 100
Chapter 3 REVIEW 104

Chapter 4 ✦ Communicating As a Scientist **106**
Lesson 1 Using Science Vocabulary 108
Lesson 2 Using Illustrations and Models 112
Lesson 3 Making and Using Graphs 116
Lesson 4 Writing Scientific Reports 122
Lesson 5 Using Technology to Communicate 126
Chapter 4 REVIEW 130

APPENDIX
Careers in Science 132
Mathematics Skills 134
Glossary 136
Index 139

TO THE STUDENT

The expression "science skills" can be misleading. Often, when people hear this expression, they picture highly trained men and women using specialized equipment to carry out difficult investigations. This picture presents a very narrow and limited view of what is meant by science skills. In the broadest sense, science skills are skills that people use every day, usually without thinking about them. Most of your everyday activities—reading, writing, moving from class to class, and rearranging the clothes in your closet—involve certain skills. These same skills, refined and applied to science and science-related areas, become science skills.

This book is divided into four broad groups of skills: planning, working, thinking, and communicating. As you can see, these are all things we do every day. Each of the broad groups is broken down into several specific skills. These skills are outlined in the table of contents on pages iii and iv. Each lesson in this book has been planned to help you to identify and define a specific skill and to use that skill in science-related activities. Included at the end of each lesson are ideas for projects and activities that you can do on your own or with one or more classmates. These activities are designed to help you use and extend the skills developed in the lesson.

By the time you are finished with this book, you will have learned to recognize certain activities as skills. You will have practiced using and perfecting these skills. Finally, you will have come to understand how everyday skills become "science skills."

CHAPTER 1

Planning As a Scientist

Solving problems is a very important part of science. The first step in problem solving is to identify the problem. Once you have identified the problem, you need to prepare a plan that will help lead you to a solution. Forming the plan will involve developing questions and thinking about possible answers to the questions. The plan will also include ways to test the answers to see which ones best answer the questions.

In this chapter, you will be developing and using skills that will help you plan better. Some of these skills are observing and interpreting what you see; organizing; reading science; asking the correct questions; predicting and hypothesizing; and planning experiments. Once developed, these skills will help you to solve the many science problems you may face.

SKILLS

- Making Observations and Interpretations ..... LESSON 1
- Using Information Sources ..... LESSON 2
- Organizing Information ..... LESSON 3
- Reading in Science ..... LESSON 4
- Asking Questions in Science ..... LESSON 5
- Making Predictions ..... LESSON 6
- Forming Hypotheses for Testing ..... LESSON 7
- Planning an Experiment ..... LESSON 8

▲ **Figure 3-1** Solar prominence

SCIENTISTS like these have spent years planning how to produce energy here on Earth that is like the Sun's energy. Fusion scientists work with a device like the one in the photo. It is called a tokamak. Because fusion reactions involve so much heat, this special device was invented to contain the energy.

Solar prominences, such as the one in the smaller photo, are produced by the type of energy scientists are trying to reproduce here on Earth. Why do you think scientists are trying to reproduce the energy?

LESSON 1 Making Observations and Interpretations

Reading for a Purpose
- How can you develop observation skills?
- How can you use observations to make interpretations?
- How can you recognize the difference between observations and interpretations?

Terms to Know
observation (ahb-zuhr-VAY-shuhn) the practice of noting and recording facts and events

interpretation (ihn-ter-pruh-TAY-shuhn) an explanation of what you have seen
observe (ahb-ZERV) to pay attention to

Observing and Interpreting

A. Look quickly at this picture. Then, cover it with something.

▲ **Figure 1-2** You probably will see more details each time you look at this picture of a fair.

1. What did you see at first glance? Make a list of all the objects you recognized.
   _____

2. Look at the picture again, but look more closely this time. What can you recognize now? _____

3. Glance at something in the classroom. Describe the item. _____
   _____

4. Now, study the same item more closely. What details can you see after a second, closer look? _____
   _____

4 | **Chapter 1** Planning As a Scientist

What you **observe** or notice about something the first time may be different from what you observe about the same object later.

B. Look at these two pictures of cats. Write what you observe about each in the chart below. Then, check the appropriate interpretation.

▲ **Figure 1-3** Note the differences between the two pictures of cats.

Observations	Interpretations (Check One)
5. Cat A _____ mouth _____ back _____	_____ quiet, relaxed _____ scared, frightened
6. Cat B _____ mouth _____ back _____	_____ quiet, relaxed _____ scared, frightened

You observed differences in the cats' behaviors. Making **observations** leads to an explanation, or **interpretation** of what you observed. You may observe an object several times before you make a definite interpretation.

C. Now, tell how your observations led to the interpretations of the pictures above.

7. Cat A _____   8. Cat B _____

Lesson 1 Making Observations and Interpretations

Practice Observing and Interpreting

Each sentence below is either an interpretation or an observation.

A. Study Figures 1-4 to 1-7 and read the statements in Items 1 to 4. For each sentence, write *O* for observation or *I* for interpretation. The first one has been done for you.

▲ **Figure 1-4** Volcano

▲ **Figure 1-5** Flat tire

▲ **Figure 1-6** Marshmallow

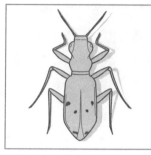
▲ **Figure 1-7** Bug

1. __O__ **a.** The volcano is erupting.

   __I__ **b.** The volcano is active.

2. _____ **a.** The tire is flat.

   _____ **b.** The tire has a leak.

3. _____ **a.** The marshmallow has changed shape.

   _____ **b.** Heat has melted the marshmallow.

4. _____ **a.** The bug is an insect.

   _____ **b.** The bug has six legs.

B. Look at the two pictures in Figure 1-8. Then, list your observations and interpretations for each picture.

Observations

5. Can A _____

6. Can B _____

Interpretations

7. Can A _____

8. Can B _____

9. How do your observations and interpretations compare with those made by others in the class?

_____

_____

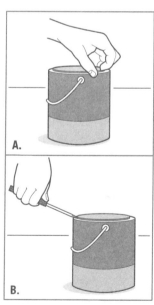
▲ **Figure 1-8** When you observe, you can then make an interpretation about what you see.

6 | Chapter 1   Planning As a Scientist

Thinking About Observing and Interpreting

Observe the pictures in Figure 1-9. Make your observations as accurate as possible. This leads to more sensible and useful interpretations.

▲ **Figure 1-9** You can observe two scenes and find similarities and differences.

1. What do you observe about these two scenes? _____
   _____

2. What interpretations can you make? _____
   _____

Extending Your Experience

Answer the following in complete sentences. Use a separate sheet of paper.

1. Using your eyes is not the only way to make observations. Make a list of other ways you can make observations.

2. Copy a color photo on a black and white copier. List all the things you can observe in the color photo that do not appear on the copy.

3. Use your ears to observe certain properties of sound. What do voices sound like inside the classroom? Outside the classroom? Two rooms away? How does distance affect sound? How do solid barriers affect sound?

4. Test your observations and interpretations of various substances. Form a group with three other students. Use the teacher's desk as a base. Position three students around the room. Have one student sit in front of the teacher's desk, another to the side and away from the desk, and a third toward the rear of the room. Use three items for your experiment, such as a bottle of vanilla extract, a jar containing half a large onion, and a bottle of perfume. Position yourself near a clock with a second hand. Open each item one at a time. Time how long it takes for each student to smell that item from his or her location.

5. A telescope extends your ability to see distant objects. Make a list of other science tools that improve your ability to observe.

LESSON 2 Using Information Sources

Reading for a Purpose
- How can you find and use information from various sources?
- Do you know the differences among the various information sources and the different kinds of information they provide?

Terms to Know

label an attached tag or highlighted section on a product that identifies the product and provides information about the product and its use

information a collection of facts or data obtained in any manner

source person or place from which information is obtained

pamphlet small, unbound booklet of printed material

Finding Information

A label is a source of information. **Labels** can show many different kinds of information. One way to find **information** about a product is to read its labels.

Label A

Pain Relief
ACETAMINOPHEN

Directions: Adults and children 12 years of age and older. Take 2 tablets every 4 to 6 hours as needed. Do not take more than 8 tablets in 24 hours, or as directed by a doctor. Do not use this product for children under 12 years.

Do NOT Use: for more than 10 days for pain unless directed by a doctor.

Stop Using and Ask a Doctor If: symptoms do not improve, new symptoms occur, pain or fever persists or gets worse, redness or swelling is present.

60 Tablets - 500 mg each

Brownseville, NJ, USA

Label B

BRIGHT WHITE BLEACH
DIRECTIONS FOR LAUNDRY USE:
Pretreat Stains and Soils: Stubborn stains may be soaked for 5 minutes in a solution of 1/4 cup Bright White Bleach to 1 gallon of cool sudsy water.
Sort Laundry by Color and Fabric: Separate whites from colors, light colors from dark ones. Avoid bleaching wool, silk, mohair, leather, spandex, non-fast colors and flame retardant cotton fabrics.
DANGER: Keep out of reach of children. See back panel for additional cautions. **STATEMENT OF PRACTICAL TREATMENT (FIRST AID): IF CONTACT WITH EYES OCCURS,** flush with water for a least 15 minutes and get prompt medical attention. **IF CONTACT WITH SKIN OCCURS,** wash with plenty of soap and water. **IF SWALLOWED,** drink large amounts of water. Do not induce vomiting. Call a physician or poison control center immediately.
Active Ingredient, Sodium Hypochlorite........6%
Inert Ingredients......94%

96 FL OZ (3QT) 2.84L Manufactured in Smithtown, NJ, USA

▲ **Figure 1-10** Examples of labels found on common household items

A. Study the labels on the products in Figure 1-10. Read the following list. Write the letter(s) of the labels that give the information listed.

_____ 1. where manufactured   _____ 5. ingredients

_____ 2. safety guidelines   _____ 6. health warning

_____ 3. volume or weight   _____ 7. first aid

_____ 4. name of product   _____ 8. directions for use

8 | Chapter 1 Planning As a Scientist

B. Carefully look again at the labels in Figure 1-10. Then, answer Items 9 to 12.

9. What is the active ingredient in the bleach? _____

10. What warning information found on the label for bleach is very important when washing clothes? _____
_____

11. What kinds of information appear on the acetaminophen bottle? _____
_____

12. What are the safety guidelines for using the acetaminophen? _____
_____

C. Look at Figure 1-11 to help you answer Items 13 and 14.

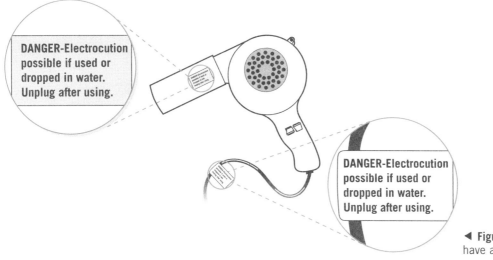

◀ **Figure 1-11** These labels have a very important purpose.

13. The warning symbol shown is actually in red and white. Explain why warning labels are in bright colors. _____
_____
_____

14. What are the precautions for using the hair dryer?
_____
_____

Lesson 2   Using Information Sources

Practice Finding Information

Some other **sources** of information include pamphlets, magazines, and the Internet. Some science pamphlets are designed to provide very specific information. The **pamphlet** may be concerned with only one aspect of science or it may group several aspects together.

Use the pictures of the different information sources in Figure 1-12 to help you answer the questions in Items 1 to 5.

▲ **Figure 1-12** You can find important information from sources such as magazines, pamphlets, and the Internet.

1. Which sources give general information about a subject or topic?

   _____

2. Where might you find very specific information about a single subject or topic?

   _____

3. In which sources will you find stories about scientific news?

   _____

4. Which sources might provide forms for you to fill out?

   _____

5. Why is it important to use different sources of information?

   _____
   _____

Thinking About Finding Information

There are many sources you can use to find information. Look at the examples of sources listed below. Think of times when you may have used some of these sources of information. Then, read the questions. Decide which of the information sources you would use to answer each question. Write the letters of the sources you would use in the spaces provided.

a. newspapers, magazines, periodicals
b. pictures or diagrams
c. maps or charts
d. family
e. container labels
f. Internet sites
g. speeches and debates
h. daily log or journal
i. personal observations
j. instruction booklet
k. reference books
l. radio or TV specials
m. interviews of experts
n. repair manuals

_____ 1. You want to go fossil collecting tomorrow. How could you find out about fossils in your area?

_____ 2. You want to read about the accident that happened yesterday at the nuclear power plant.

_____ 3. Where would you look to find out about your favorite scientist?

_____ 4. How could you find out about using complex laboratory equipment?

_____ 5. You want to write a paper on a science topic of interest to you. List the sources you would use to research the topic.

_____ 6. Where could you find out about the ingredients of a specific product?

_____ 7. Where would you find information for a report on how lasers work?

_____ 8. Where do you find the latitude of an island?

_____ 9. You want to find out how to fix a faulty piece of equipment.

_____ 10. How often during one day do you speak on a telephone?

Extending Your Experience

Answer the following in complete sentences. Use a separate sheet of paper.

1. What sources of information can you use to learn how a microscope works?

2. Select a topic having to do with Earth's history. List at least four sources of information you would use to research your topic. Briefly describe the information each source would provide.

3. Find pictures of different machines in action. For each picture, write an observation and an interpretation. Also list the sources of information you would use to support each observation and interpretation.

4. Choose any object you wish and design a label for it. Review Lesson 2, Using Information Sources, for the kinds of information you might include on the label.

Lesson 2 Using Information Sources

LESSON 3 Organizing Information

Reading for a Purpose
- How can you organize information in a chart?
- How can you use information from a chart?

Term to Know
chart a presentation of information in a visual, easy-to-read format

Information From a Picture

Have you ever heard the expression "A picture is worth a thousand words"? Often you can get as much information from looking at pictures as you can from reading words.

You can organize the information you get from a picture. One way to do this is to put the information in a chart. A **chart** is an easy way to organize information. The data you collect can be displayed in an orderly arrangement. Sometimes, organizing information in a chart can help you learn more from the information. For example, the chart might show some important relationships.

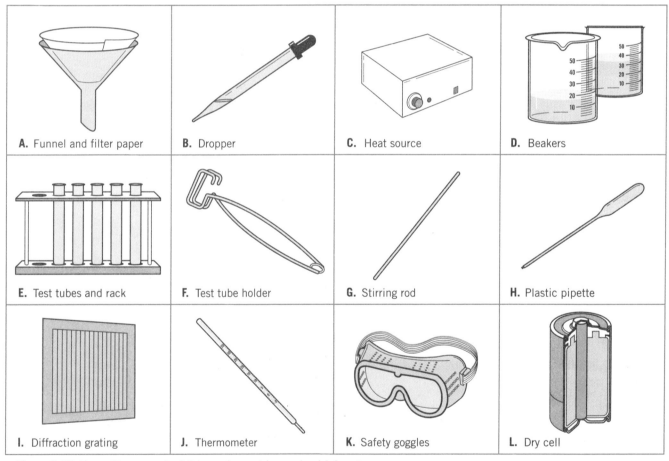

▲ Figure 1-13 You will see most of this equipment in a school laboratory.

12 | Chapter 1 Planning As a Scientist

A. Use the information in Figure 1-13 to fill in the chart in Figure 1-14.

Uses of Laboratory Equipment				
Used to Measure	Used to Hold Equipment	Used to Transfer Substances	Used For Safety	Used to Hold Substances
thermometer				
plastic pipette				
beakers				

▲ Figure 1-14 You can organize your information in a chart.

B. Use the space below to make a chart that organizes this same equipment in a different way, such as items made from glass and items made from metal. You may think of other ways to organize this equipment.

C. Use the information in Figures 1-13 and 1-14 to answer the following questions.

 1. What are some different uses for laboratory equipment? _____

 2. In which column did you list the most items? _____

 3. Why do you think organizing is important to scientists? _____

Lesson 3 Organizing Information | 13

Practice Organizing Information

Machines make work easier. Different kinds of machines do different kinds of jobs. Simple machines make up parts of compound machines.

Figure 1-15 ▶ Three simple machines

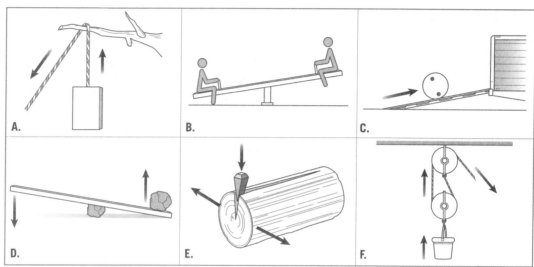

▲ Figure 1-16a You can organize types of machinery by setting up a chart much like this one.

A. Look at the three different kinds of simple machines shown in Figure 1-15. Then fill in Figure 1-16b by writing letters of each drawing in the column that names that simple machine as shown in Figure 1-16a.

Types of Simple Machines		
Pulley	Lever	Inclined Plane

◀ Figure 1-16b Information from one chart can help you fill out information in another chart.

B. List at least two examples of each simple machine. Use the information in Figures 1-15, 1-16a, and 1-16b.

1. levers _____

2. pulleys _____

3. inclined planes _____

14 | Chapter 1 Planning As a Scientist

Thinking About Organizing Information

Read and answer the following questions.

1. On page 13, you used a chart to organize laboratory equipment. What did you do to organize this information? Place a check (✔) by the things that you did.

   _____ a. identified uses

   _____ b. wrote names of items on the chart

   _____ c. looked for items that were alike

   _____ d. counted items

   _____ e. identified colors

   _____ f. identified sizes

   _____ g. sorted items into groups

   _____ h. looked for items that were different

   _____ i. identified what items were made of

   _____ j. identified items by weight

2. How can organizing information help you improve your grades in school?

   _____

   _____

   _____

Extending Your Experience

Answer the following in complete sentences. Use a separate sheet of paper.

1. Think of several activities in which you participate that require organization. How are they organized? Do they each have a different type of organization?
2. Write down your name and address and think what each part of it means. Describe how the name and address are organized.
3. Design some cabinets for storing laboratory equipment. Label what types of equipment will be stored and where it will be stored. Explain how you would organize the equipment so that items can be found easily.
4. Look in the card catalog at the library for information about simple machines, such as levers, gears, and inclined planes. How are the books under that topic arranged? Look up one of the books you found in the card catalog. How is information organized within the book?
5. Newspapers provide a lot of current information. Read a newspaper and tell the ways it organizes the information it provides.

LESSON 4 Reading in Science

Reading for a Purpose
- How can you get information from photographs?
- How can you get information from diagrams?
- How can you use information from text?
- How can you relate a picture to the information in a caption?

Terms to Know
diagram a drawing that clearly shows how something is arranged

caption (KAP-shuhn) a short description under or beside a photograph, diagram, map, or illustration

Reading Photographs, Diagrams, and Captions

An electromagnet uses electric current to make a strong magnet. When the electricity is off, the magnetism disappears.

You can discover many things by looking closely at a photograph.

A. Study the photograph of an electromagnet at work in Figure 1-17. Then, answer Items 1 to 3.

1. Is the electric current on or off in the electromagnet? Explain your answer.

   _____

   _____

2. How can the electromagnet move metal from one place to another?

   _____

   _____

3. What other observations can you make from this picture?

   _____

   _____

▲ Figure 1-17 This crane has an electromagnet.

Sometimes you can get more information from a diagram than from a photograph. A **diagram** is a picture drawn to show clearly how something is arranged. For example, a diagram of an electromagnet might show how electricity creates a magnetic field (something the photograph in Figure 1-17 cannot do). A diagram of a volcano, like the one on page 17, can show the internal structure that is not visible in a photograph of the same volcano. Often, diagrams are simple drawings of the real thing. They help the reader identify important parts or processes.

Chapter 1 Planning As a Scientist

B. In Figure 1-18, compare the picture of a volcano to the diagram. Then, answer Items 4 to 6.

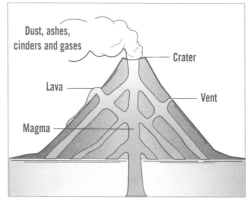

▲ **Figure 1-18** A photograph and a simple drawing can help you identify important parts of a volcano.

4. Which shows better what a volcano actually looks like? _____

5. Which shows better how a volcano works? _____

6. Which shows better the parts of a volcano? _____

When you read a book, both text and pictures help you to understand. Often, a picture will have a caption, like those you see on the pages of this book. A **caption** is a short description under or beside the picture.

C. Look at the pictures and labels in Figure 1-19 and read the captions. Then, read Items 7 to 10. In the spaces provided, write *P* if you find the answer by looking at the pictures and labels. Write *C* if you find the answer by reading the captions.

▲ **Figure 1-19** All mixtures are made up of two or more kinds of matter. The pictures show different mixtures before and after being separated by physical means.

_____ 7. What is true about all mixtures?

_____ 8. How many mixtures are shown in the pictures?

_____ 9. How do the mixtures differ?

_____ 10. How do you think the materials in the liquid mixture were separated?

Lesson 4 Reading in Science   17

Practice Reading in Science

Study the pictures and read the labels and caption in Figure 1-20. Then, read the text that follows and complete Items 1 to 4 below. Write the correct magnet next to the statement.

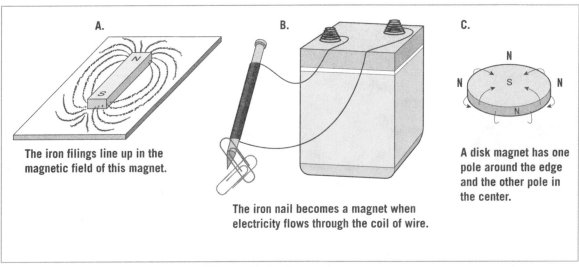

▲ Figure 1-20 The pictures show three different kinds of magnets.

There are different types of magnets. All magnets attract iron-bearing metals and objects with electrical charges. Some bar magnets form when melted iron cools in a strong magnetic field and the iron molecules line up with the magnetic field.

Electromagnets become magnetic when there is an electrical current present. Electromagnets can be weak or strong. The greater the number of turns of wire and the stronger the electrical current, the stronger the electromagnet.

Disk magnets are shaped like coins. These magnets are used in radio and stereo speakers.

1. Poles are at the ends of the magnet. _____

2. Magnetism can be turned on and off. _____

3. These are used in stereo speakers. _____

4. These attract iron-bearing metals. _____

Thinking About Reading in Science

Science books, magazines, Web sites, and other written materials combine text, photographs, diagrams, and charts to give information.

A. Describe ways that each of the following can give information in a way that the others cannot. You may use examples to help you explain.

1. color photograph _____

2. labeled diagram _____

3. chart _____

18 | Chapter 1 Planning As a Scientist

B. Figure 1-21 shows a series of illustrations. Read them from left to right. Describe what these pictures tell you in the spaces provided.

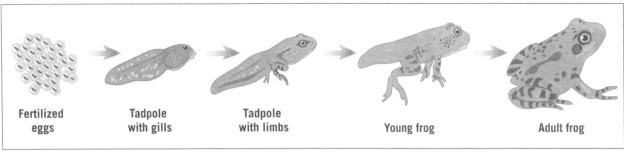

▲ Figure 1-21 Pictures can tell a story.

4. _____

Extending Your Experience

Answer the following in complete sentences. Use a separate sheet of paper.

1. Take a series of photographs of objects that change in response to heat. Examples might include a melting ice cube and the puddle that forms as it melts. Another might be salt water evaporating, leaving the salt behind. Display your photograph with captions and write a brief description.

2. Pick an object that you know a lot about, such as a bicycle or a desk. Draw a diagram that shows the important parts. Label each important part. Draw a straight line from each label to the part it names. Then, briefly describe the object.

3. You have learned that reading can be more than getting information from written text. You may have to read numbers, musical notes, or body language for meaning at some time. Work with a partner and list five other things you can read.

4. Pick four pictures in your science textbook. Write a new caption for each picture. Each new caption should mention something different from the old caption.

5. Find an article in a newspaper about a science topic but without pictures. What photographs would help you understand the article better? Who would take those photographs? What diagrams would have been useful?

LESSON 5 Asking Questions in Science

Reading for a Purpose
- How can you improve your skills in asking questions?
- How do you identify different ways to answer questions?

Terms to Know

general broad
specific narrow
experiment an investigation or test
expert a person who knows a lot about a certain topic
dictionary a book that gives pronunciations and definitions
encyclopedia a book that gives general and specific information on many topics
reference book any book that provides information

Asking and Answering Questions

Asking questions is important in science. By asking questions, a scientist can find interesting subjects to study. A question map is one way to think of questions. You can make a question map by writing a topic, or subject, on a sheet of paper. Around the name of your topic, write some related questions. You will find that one question will lead to another question. You can draw lines between the questions. Below is an example of a question map about heat.

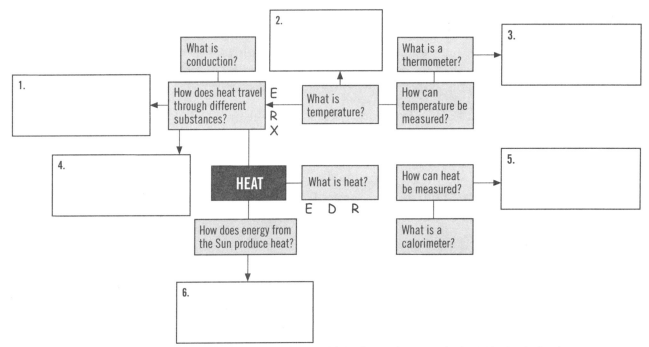

▲ Figure 1-22 A question map can help you think of questions to ask about the topic *heat*.

A. Use the questions on the map in Figure 1-22 to lead you to other questions. Write them in the blanks on the map. Add more questions as you think of them. Draw lines between the related questions.

20 | Chapter 1 Planning As a Scientist

Some questions are **general**, or broad. Other questions are **specific**, or narrow. In science, specific questions can be answered through planned experiments and observations. By answering questions, scientists gain understanding and learn facts. When an **experiment**, or investigation, gives a clear answer, a scientist may write a book or article on the topic. More often, answers to one question lead a scientist to ask another question. There are always new questions to be asked. Figure 1-23 shows a general question that has been broken into specific questions.

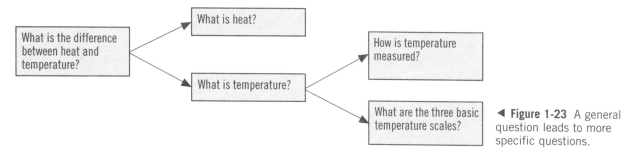

◀ **Figure 1-23** A general question leads to more specific questions.

B. Answer Items 7 to 9 in the spaces provided.

7. Add some more specific questions to the general question in Figure 1-24.

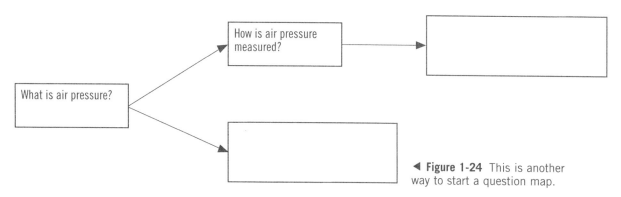

◀ **Figure 1-24** This is another way to start a question map.

8. Look at Figure 1-22 on page 20. Find a question that is general. Write it here.

_____

_____

9. Look again at Figure 1-22 on page 20. Find a specific question. Write it here.

_____

_____

There are several ways you can find answers to questions. **Experts** are people who know a lot about a certain topic. They can give you information if you ask them.

Dictionaries give definitions, often using examples. They also can tell you how to pronounce a word. An **encyclopedia** or **reference book** has both general and specific information about most topics. An **experiment**, or investigation, is a way to find and test answers to a question or problem. An experiment can give you new information. It can also give you first-hand experience.

Lesson 5 Asking Questions in Science | 21

C. Answer Items 10 and 11 in the space provided.

10. What other ways do you know to find answers to questions?

   _____

11. Look again at Figure 1-22 on page 20. Label the questions on the map to show how you could get answers. Under each question box write *E* if you would ask an *expert*, *D* if you would look in a *dictionary*, *R* if you would use a *reference book*, or *X* if you would try an *experiment*. You might find that you can use more than one label for some questions. Your choice of labels may be different from other students' choices. Think of how you would explain your label. Two questions have already been labeled for you.

   _____

Practice Asking and Answering Questions

Use the space on this page to make your own question map. Begin by writing the name of a science topic you are now studying. If you cannot think of a topic, you may choose one of those in the box.

| plants | photosynthesis | animal cells | genetics |

Use arrows to lead from general questions to more specific questions. Label the ways you could find answers to your questions by writing *E*, *D*, *R*, or *X* as you did for the questions on page 20.

Thinking About Questions in Science

Asking questions is one important way to get answers in science. Other ways you can get answers are to observe, experiment, and collect information. You might need to do only one or two of these to find an answer. Sometimes you need to do them all. Often, steps must be repeated. Below is one path you can use to get answers. There are many other paths. Notice that you may start at any step.

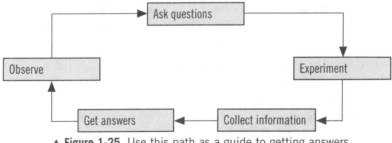

▲ **Figure 1-25** Use this path as a guide to getting answers.

Below are some questions about Figure 1-26. For each question, list the steps in the path that you would take to get the answer. Would you observe first? Would you collect information first? What would you do second? When would you experiment? Add new steps if they are needed to gather information.

1. What kinds of energy are being shown?

   _____

   _____

   _____

2. How is the magnifying glass used?

   _____

   _____

   _____

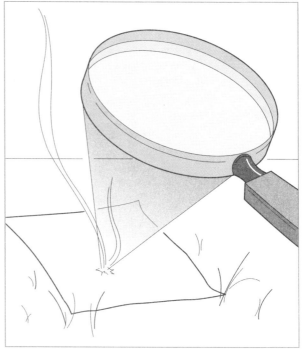

▲ Figure 1-26 A magnifying glass can be used to start a fire.

Extending Your Experience

Answer the following in complete sentences. Use a separate sheet of paper.

1. List five facts, or answers, about electricity and magnetism. Write a question for each one that might have led to that fact or answer. Read your answers to the class. Then, ask them to tell you what they think the question might have been.

2. Form a group with five class members to hold an *ask-off* competition. Choose a topic in the science field that you are presently studying. Have each one in the group take turns asking a question about that topic. The last person who is able to ask a question on that topic wins.

3. List the steps in the path you would take to answer these questions: Why do different objects have different colors? How does a waterfall form?

4. Use an encyclopedia, the Internet, or other reference to find out what each of these science experts does: animal breeder, flight engineer, marine biologist, and geologist.

5. Pick three science words you have trouble pronouncing. Look them up in a dictionary. Most dictionaries have a pronunciation guide. The guide will help you decide how to pronounce the words. Write the definitions and the pronunciation of the words. The pronunciation of each word is found in parentheses next to the entry.

Lesson 5 Asking Questions in Science

LESSON 6 | Making Predictions

Reading for a Purpose
- Why make predictions?
- What is the difference between an inference and a prediction?

Terms to Know

infer to form a conclusion based on evidence; explain or interpret an observation

prediction a statement made ahead of time about what you think might happen

dominant gene a gene whose trait always shows itself

recessive gene a gene whose trait is hidden when the dominant gene is present

Punnett square a chart that shows possible gene combinations

Making Inferences

To **infer** is to form a conclusion about something. However, when we make an inference, we base it on observations, prior knowledge, and past experiences. Inferences are not just guesses, nor are they facts. They should be supported by evidence. Inferences can be tested by experiments. An inference is an explanation or interpretation of our observations. We infer from things in our everyday lives.

Scientists make inferences as they investigate. Then, they form hypotheses, or possible answers, to test and investigate.

A. Look at the two columns in Figure 1-27. You may have made some of these observations today. Fill in the *Inferences* column with two inferences for each observation. The first one has been done for you.

Observations	Inferences
Many classes are not well attended.	1. Students are on a class trip. 2.
There was no mail in the mailbox when you arrived home.	3. 4.
It is dark outside, but it is noon.	5. 6.

Figure 1-27 ▶
Observations can lead to inferences.

B. Now, describe the observations, prior knowledge, and past experiences that you used to make the inferences you made above.

_____

_____

Making Predictions

After a scientist has made an inference about a topic, he or she will next make a prediction. A **prediction** is stating ahead of time what you think might happen. This is an important step before designing and performing an experiment. By making predictions, scientists have a good idea about an experiment's success and outcome.

When attempting to formulate a prediction, there are two questions you can ask: *What will happen next if this occurs? If I do this, what will happen?*

A. Look at the Items in Figure 1-28. Then, answer Items 1 and 2 in the spaces provided.

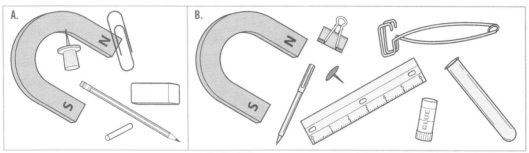

▲ Figure 1-28 You can make inferences and predictions based on observations.

1. Use Picture A in Figure 1-28 to make an inference about objects that are attracted by a magnet. _____

   _____

2. Based on your observations and inferences for Picture A, predict what will happen when you bring the magnet near each of the items in Picture B.

   _____

   _____

Even though inferences and predictions are both made based on observations or prior knowledge, they are different. Inferring is a conclusion, explaining why something occurred; predicting is stating what you expect will happen in the future.

Lesson 6 Making Predictions | 25

B. Study the table of barometric pressure readings for four consecutive days in Figure 1-29. Then, look at the sequence of cloud formations for the first three days in Figure 1-30. Answer Items 3 and 4 in the spaces provided.

Barometric Pressure Readings				
Day	Monday	Tuesday	Wednesday	Thursday
Reading	30.06	29.90	29.50	29.35

◀ **Figure 1-29** You can make an inference from information in a chart.

▲ **Figure 1-30** You can make predictions based on observations of photographs.

3. What inferences can you make from the information in the chart and the photographs?

   _____

4. Predict what is likely to happen on Thursday based on what you inferred.

   _____

Practice Making Predictions

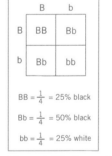

▲ **Figure 1-31** Punnett squares can help predict gene combinations in offspring.

One field of science in which scientists often make inferences and predictions is genetics. In the field of genetics and reproduction, there are dominant and recessive genes. A **dominant gene** is one whose trait always shows itself when present. A **recessive gene** is one whose trait is hidden when the dominant gene is present. **Punnett squares** are charts that show possible gene combinations, using a capital letter for the dominant gene and a lowercase letter for the recessive gene. Scientists make use of these when predicting possible outcomes of genetic combinations.

Figure 1-31 is one example of a Punnett square for two hybrid black guinea pig parents (Bb), each with a recessive gene (b) for white. From the information, you can make inferences and predictions.

A. Complete the Punnett square in Figure 1-32 to show what happens when you cross a hybrid black guinea pig (Bb) with a white one (bb). Then, predict the possible colors the offspring will have.

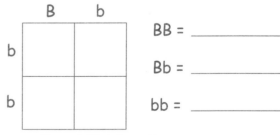

▲ **Figure 1-32** A Punnett square.

BB = _____

Bb = _____

bb = _____

26 | Chapter 1 Planning As a Scientist

When you buy flower seeds, you look at the colors of the flowers on the package to help you design your garden. For one part of the garden, you select a package of seeds that shows all red flowers. You know that the color red (R) is a dominant trait in these flowers. However, the label on the package tells you that all of the seeds are from hybrid plants and have a recessive gene for white (r). Will all of the flowers produced by the seeds in this package be red?

B. Create a Punnett square to predict what you may see in the garden when the flowers bloom.

1. What color will most of the flowers be? _____

2. What percent of the flowers will not be red? What color will they be?
_____

	R	r
R		
r		

▲ **Figure 1-33** A Punnett square showing possible flower color combinations

Thinking About Predicting

In the science laboratory, you are instructed to mix 5 grams of salt and 50 mL of water at 10°C.

1. Describe what you already know about salt, water, and the combining of the two substances. _____
_____

2. Then, predict what would likely occur if you added 15 more grams of salt to the water. _____
_____

3. How might you successfully dissolve all the salt in the water? _____
_____

Extending Your Experience

Answer the following in complete sentences. Use a separate sheet of paper.

1. Follow the records of two sports teams of your choosing. Then, make a prediction of which team will win when they play each other. Check the results of the game to see if your prediction was correct.

2. Find the stock market report in the newspaper. Choose a stock and follow its progress for two weeks. Would you buy this stock? Explain why you chose this particular stock. Use inferences and predictions in your explanation.

Lesson 6 Making Predictions | 27

LESSON 7 Forming Hypotheses for Testing

Reading for a Purpose
- How can you recognize the importance of good hypotheses?
- How do you form a hypothesis?

Term to Know
hypothesis (hy-PAHTH-uh-sihs) a possible answer to a scientific question based on information

Forming Hypotheses

Figure 1-34 is a photograph of a natural stone arch. The arch has not always been there. It was formed over thousands of years. What do you think caused the arch to form?

Scientists who study the land make careful observations. In this case, many arches were studied to see how they are alike. Scientists think of questions and then try to find answers. Answering questions involves observing, collecting information, and experimenting.

Scientists think this arch was formed as weak rock broke down and was carried away. This arch is left for now. Someday, it too will break down and be carried away. Thus, scientists have answered the question, what caused this arch to be formed?

To help answer a question, a scientist forms a statement called a hypothesis. A **hypothesis** is a possible answer to a scientific question based on information. Hypotheses are formed so that a scientist has something to test—to prove or disprove.

▲ **Figure 1-34** Scientists have hypothesized about the formation of this arch.

Here is an example of a question and two hypotheses:

Question: What caused the natural arch to be formed?

Hypothesis I: An earthquake caused the arch to be formed.

Hypothesis II: The wearing away of weakened rock within a section of stronger rock caused the arch to be formed.

Hypotheses are formed after observing, comparing, and thinking. A hypothesis is not a wild guess. It is a thoughtful statement that a scientist hopes to prove correct. After testing and experimenting, scientists proved Hypothesis I to be incorrect. They proved Hypothesis II to be correct.

A. Study questions 1 to 3 in the chart. Then, read the statements below the chart. Which statements are hypotheses? Which statements are ways of testing hypotheses? For each question, there is one hypothesis and one test method. Write the letters in the correct column in the chart for each question. One answer has been done for you.

Question	Hypothesis	Test Method
1. How long does it take the Moon to make one revolution around Earth?	c	
2. In your area, which month has the most rainfall?		
3. Which is harder, quartz or window glass?		

a. Check records to find the rainfall each month for several years.

b. Try to scratch window glass with a piece of quartz.

c. It takes the Moon about 27 days to make one revolution around Earth.

d. Quartz is harder than the window glass.

e. Count the days between full moon phases for several months.

f. April is the rainiest month.

The best hypothesis is one that can be tested. A hypothesis can be tested when you can either prove or disprove it with hard evidence. It is also important to write a hypothesis clearly. Others might want to test your hypothesis. They should know exactly what you mean.

B. Read the question and hypothesis. Then, answer Items 5 and 6.

Question: How is an ant different from a fly?
Hypothesis: Ants are smaller than flies.

4. The hypothesis just stated would be hard to test because of the word *smaller*.

   Explain why. _____

   _____

5. How could you test ants and flies to show their differences?

   _____

   _____

C. For Items 6 and 7, compare the two statements. Check (✓) the hypothesis that can be tested.

6. How are sandstone and quartzite alike?

   _____ a. Sandstone and quartzite are both pretty.

   _____ b. Sandstone and quartzite are both composed of quartz.

Lesson 7 Forming Hypotheses for Testing

7. How does the temperature of water affect how much sugar can be dissolved in it?

   _____ a. Sugar will dissolve faster in water that is being stirred.

   _____ b. Sugar dissolves better in warm water than in cold water.

Practice Forming and Testing Hypotheses

You can form a hypothesis about almost any topic. However, a hypothesis must be testable. Always try to state your hypothesis as simply and clearly as possible.

A. Read the hypotheses in Items 1 to 3. Then, describe how you could test each hypothesis in the spaces provided.

1. Every day, the Moon rises about 50 minutes later than the day before.

   _____

2. Sound travels faster through water than through air.

   _____

3. The mineral gypsum always produces a white streak, no matter what the color of the mineral's surface.

   _____

B. Answer Items 4 and 5 in the spaces provided.

4. a. Write a hypothesis about the relationship between the weight of an object and the rate at which it falls. _____

   _____

   b. How could your hypothesis be tested? _____

   _____

5. Write a hypothesis to explain the relationship between air pressure and altitude.

   _____

Thinking About Hypotheses

Testing a hypothesis may be difficult if the wording of the statement is not very clear. Therefore, hypotheses should be stated clearly. They also should be testable. If something can be measured, it is tested more easily. Most things have recognizable traits, such as size and color. The following are three reliable methods of testing.

- **Measuring** In many cases, the easiest and most reliable way to test something is to measure it. However, many hypotheses deal with things or ideas that cannot be measured directly.

30 | Chapter 1 Planning As a Scientist

- **Comparing** Often, something can be tested by comparing it with something else that is known or familiar.
- **Personal Opinion** Personal opinion is the least reliable type of test. Ask 20 people for their personal opinions about something and you are apt to get 20 different answers.

A. Look at the picture of a fossil in Figure 1-35. Then answer Items 1 and 2 in the spaces provided.

1. Each of the following entries relates to the fossil. Beside each item, write *M* for those that can be *measured*; *C* for those entries that must be *compared* with something; and *P* if the item requires a *personal opinion*.

▲ **Figure 1-35** Fossil of fish, *Priscacara oxyprion*, found in Green River, Wyoming

　　　　 a. length of the fossil

　　　　 b. altitude at which it was found

　　　　 c. beauty of the rock in which it was found

　　　　 d. age of the fossil

　　　　 e. how easy it was to climb the mountain

　　　　 f. type of animal it was

　　　　 g. type of rock in which it was found

　　　　 h. scientific value of the fossil

2. You are asked to form a hypothesis related to the dissolving of salt in water. Describe three measurable or observable traits that you could use.

_____

_____

Extending Your Experience

Answer the following in complete sentences. Use a separate sheet of paper.

1. If a hypothesis proves incorrect, should the scientist feel a sense of failure? Do you think anything good could come from an incorrect hypothesis? Explain.

2. Mechanics make observations about the condition of automobiles. They have to make hypotheses about things that they cannot see. To test their hypotheses, they perform a variety of tests using many different kinds of equipment, including stroboscopes, wheel balancers, tachometers, and others. Find information about one of these tests. Describe the test and what information can be gained by doing the test.

3. Search the newspaper for an event that concerns volcanoes, hurricanes, or other science topics. Suggest one or two hypotheses that could explain why the event occurred. For each hypothesis, suggest a way it could be tested.

LESSON 8 Planning an Experiment

Reading for a Purpose
- How do you identify variables and controls in an experiment?
- Why use variables and controls to plan an experiment?

Terms to Know
variable some thing that may change, or vary
control a variable that does not change
control experiment an experiment in which a variable is held constant

Planning Variables in an Experiment

While writing a report, Lauren knocked some paper off her desk. A sheet of notebook paper fluttered slowly to the floor. A crumpled ball of notebook paper fell quickly to the floor. Lauren decided to set up an experiment to find out why these two pieces of paper fell in different ways and at different speeds.

The first thing that Lauren needed to do was to time how quickly each piece of paper fell. She gathered a stopwatch, some sheets of notebook paper, and a stepstool. She asked her lab partner Kamaria to help. Lauren asked her to stand on the stepstool and stretch out her right arm. For Part 1 of the experiment, shown in Figure 1-36, she handed Kamaria a sheet of notebook paper. When Lauren said, "Go," Kamaria dropped the sheet as she started the watch. She stopped the watch when the paper hit the floor. The watch stopped at 3 seconds.

▲ Figure 1-36 Kamaria stood on the stool first.

For Part 2, Kamaria measured time and Lauren dropped the paper. She wadded up a sheet of notebook paper and raised it as high as she could in her right hand. When Kamaria said, "Go," Lauren released the paper. Kamaria stopped the watch the moment the paper hit the floor. The time was 3 seconds.

Lauren was surprised that the ball of paper and the sheet of paper both took 3 seconds to fall. She remembered that the papers she knocked off the desk clearly fell at different speeds. Lauren wondered what made the difference between the results of the experiment and what she saw when the papers fell naturally.

A number of things may have made the difference. They are things that varied, or changed. Thus, they are called **variables**. Lauren thought for a while. Then, she listed the variables below.

A. **Study the list of variables below. Then, write *S* beside the variables that were the same for Part 1 and Part 2 of the experiment. Write *X* beside the variables that changed from Part 1 to Part 2.**

  _____ 1. use of stepstool   _____ 3. person who timed the paper fall   _____ 5. shape of paper

  _____ 2. height of hand above floor   _____ 4. watch used to measure falls   _____ 6. kind of paper

Lauren wanted to find out why the times in Part 1 and Part 2 were the same in the experiment. She decided to test, or experiment, how each variable may have affected the results.

B. **Re-read carefully the experiment Lauren performed to test the variables. Then, answer Items 7 to 12 in the spaces provided.**

  7. Review your responses to Part A above. Then, list the variables that were different between Part 1 and Part 2 of the experiment?

  _____

  8. Lauren wants to find out which variable or variables caused the papers to take the same time to fall. To do this, she must change one variable at a time and keep all the other variables the same. A variable that does not change is called a control variable, or **control**. Which variables in the list were controls?

  _____

  9. Lauren wants to make sure that each paper falls the same distance. To do this, which variable should she make a control?

  _____

  10. How can Lauren make sure the watch is started and stopped the same way each time?

  _____

  11. What is the only variable Lauren wanted to test? _____

  12. Describe how Lauren can repeat the experiment to get an accurate time for each paper as it falls.

  _____

  _____

Lesson 8 Planning an Experiment

Practice Planning an Experiment

Often, a scientist will test a variable by conducting two or more experiments at the same time. One experiment includes the variable to be tested. The other experiment does not include that variable. Everything else is the same. The experiment that holds the variable constant is called the **control experiment**.

For instance, a scientist may test this hypothesis: *Contact with air causes moisture to evaporate.* To test the hypothesis, the scientist places a damp sponge in an open container. As a control, an identical sponge dampened with the same amount of water is placed in an identical container and covered with a lid. Later, the two sponges are squeezed and the amount of water each holds is compared.

Here is a question that can be answered by conducting an experiment: Do different colors absorb different amounts of heat from sunlight? Look at Figure 1-37 for possible materials to use while conducting this experiment.

White paper   Black paper   Gray paper   Thermometers

▲ **Figure 1-37** These materials can be used to test absorption of heat from sunlight.

A. Use Figure 1-37 to help you answer the following questions.

1. What are some variables that might be involved? _____

2. Would a control experiment be helpful? Explain. _____

3. What variables would you test first? _____

4. What would be your hypothesis? _____

5. What would your controls be? _____

B. In the space below, describe how you would set up your experiment. You may use words, drawings, or both.

Thinking About Planning an Experiment

Remember these steps when you want to test a hypothesis by conducting an experiment.

Step 1 Plan carefully. Consider all possibilities. Think about every step.

Step 2 Before you start, talk it over with a teacher or a friend. Listen to that person's ideas.

Step 3 Consider one variable at a time.

Step 4 List all control variables. Make sure each stays the same throughout the experiment.

Step 5 Use control experiments where appropriate.

Now, list two other things you should consider or do before beginning an experiment.

_____

_____

Extending Your Experience

Answer the following in complete sentences. Use a separate sheet of paper.

1. Perform the experiment you planned on pages 34 and 35. You may want to work with friends. Because the experiment could take some time, make sure someone has responsibility for the experiment at all times.

2. Scientists cannot yet travel to the deepest parts of the oceans to perform experiments. Instead, they try to match deep-ocean conditions in the laboratory. Would experiments performed in the laboratory be just as good as the same experiments performed in the ocean? Would some variables be different?

3. Often, a scientist cannot control all the variables in an experiment. This is especially true for experiments performed outdoors. When the results are reported, should the scientist include information that some variables could not be controlled? Why or why not?

4. Physicists have performed many scientific experiments in a vacuum, using containers from which all air has been removed. Why? What variable or variables can best be tested in a vacuum?

5. Astronauts have performed many scientific experiments during space flights. Why? What variable or variables can be tested in space that cannot be tested on Earth?

Lesson 8 Planning an Experiment

Chapter 1 Review

Concept Review

A. Study the picture of the ocean shore at low tide in Figure 1-38. Write *I* if it is an interpretation or *O* if it is an observation.

_____ 1. This shoreline has no sandy beach.

_____ 2. No plants grow on the rocks.

_____ 3. The water here is very deep.

_____ 4. Waves have eroded the rocks.

_____ 5. This would be a dangerous place to swim.

_____ 6. The rocks will be covered by water at high tide.

▲ Figure 1-38 The ocean at low tide

B. Draw a line from each question to the source where you would most likely find the answer.

Questions	Sources
7. How deep is the ocean?	dictionary
8. What happens to the environment surrounding a polluted body of water?	labeled figure
	encyclopedia
9. How can you identify a type of frog?	TV or Internet
10. What are the landforms in your state?	field guide to amphibians
11. How do you pronounce *photosynthesis*?	expert
12. What are the parts of a dry cell battery?	topographic map

C. For Items 13 and 14, check (✓) the statement that would make the better hypothesis in each pair.

13. _____ Crickets chirp only at night.

_____ Crickets chirp faster in warm weather than in cold weather.

14. _____ Sound can travel through water.

_____ Sound is a form of energy.

36 | Chapter 1 Planning As a Scientist

D. Make a chart to organize the following words in an understandable way. Make up headings for each column. Use a separate sheet of paper

ocean	small	pond	thunderstorm	rainstorm
hill	lake	large	mountain	mound
meteor	star	planet	medium	tornado

E. Explain why a control experiment can be helpful. _____

_____

Vocabulary Review

Complete Items 1-12. Then, use the boxed letters to find the clue word in Item 13. Some letters have been filled in for you.

1. a person who knows a great deal about a subject __ x __ __ __ __

2. a book of definitions __ __ c __ __ __ __ __ __ __

3. where you get information __ __ __ __ e __ __

4. a possible answer to a scientific question
   __ __ __ __ __ __ __ __ s __

5. a short description under a picture __ __ p __ __ __ __

6. a variable that stays the same __ __ n __ __ __ __ __

7. something noticed or seen __ __ s __ __ __ __ __ __ __ __

8. an explanation of something
   __ i __ __ __ __ __ __ __ __ __ __ __

9. a test of a hypothesis __ __ __ __ r __ __ __ __ __

10. a drawing that clearly shows parts __ __ __ __ __ __ __ m

11. part of an experiment that can change __ __ __ __ __ __ __ e

12. a way to organize information __ h __ __ __

13. a book or set of books containing information about almost any topic

    __ __ __ __ __ __ __ __ __ __ __ __ __

Chapter 1 Review | 37

CHAPTER 2

Working As a Scientist

In Chapter 1, you learned about the ways that scientists plan before starting investigations. Now, you will read about the skills scientists apply as they are working on investigations.

Scientists around the world use the metric system as their standard of measurement, which you will study in this chapter. Measuring is one of the skills scientists use most often. Some measurements may be estimated. Others must be made directly and accurately. There are many tools that help make measuring easier and more precise. A scientist must know what tools to use and how to use them.

In this chapter, you will also learn about estimating, such as when to use an estimate instead of an accurate measurement. You will also learn about taking samples, collecting information, recording observations carefully, and keeping safe in the laboratory.

As scientists do their work, they must collect and carefully record data. If information is not recorded correctly, much hard work can be lost forever. Scientists must also work safely to avoid causing accidents or harm to themselves and others. Applying these skills correctly makes a scientist's work factual, accurate, and meaningful.

SKILLS

- **Measuring** .... LESSON 1
- **Estimating Measurements** .... LESSON 2
- **Using the Metric System** .... LESSON 3
- **Practicing Safety in the Laboratory** .... LESSON 4
- **Working in the Laboratory** .... LESSON 5
- **Sampling** .... LESSON 6
- **Recording** .... LESSON 7

VOLCANOLOGISTS like these take lava samples from a volcano in Hawaii. Once the samples are taken, the scientists will use them for many tests, such as mineral content and crystal size. Then, the volcanologists will create careful records of their findings. What other information might be obtained from lava samples? Notice the clothing they are wearing while working on the volcano. Why do you think they are dressed this way?

The smaller photo shows a seismograph recording data about earthquakes. How might this tool help in the study of volcanoes?

▲ Figure 2-1 Seismograph

LESSON 1 Measuring

Reading for a Purpose
- What are some types of measurements made by scientists?
- What is the importance of accurate measuring?

Terms to Know

measure to compare an unknown value with a known value using standard units

compare to tell how things are alike and different

length the distance from one point to another

meter the basic unit of length or distance

volume the amount of space an object takes up

liter the basic unit of liquid volume

mass the amount of matter in something

gram the basic unit of mass

temperature the measure of how hot an object is

degree Celsius (SEHL-see-us) the basic unit of temperature

scale the series of lines marked on a measuring tool

Information From Measuring

Scientists gather information by measuring. To **measure** is to **compare** an unknown value with a known value using standard units. Scientists measure length, volume, mass, and temperature. The result of measuring something is called a measurement.

Length, volume, mass, and temperature are those measurements most often made. **Length** is the distance from one point to another. The basic unit of length is the **meter** (m). **Volume** is the amount of space an object takes up. You can measure the space a solid object takes up or how much liquid or gas a container holds. The basic unit for the volume of a solid is the cubic meter (m^3). The basic unit of liquid or gas volume is the **liter** (L). **Mass** is the amount of matter in something. Mass is not the same as weight, but you can normally measure mass by weighing something. The basic unit of mass is the **gram** (g). **Temperature** is a measure of how hot an object is. It is measured in degrees. The basic unit of temperature is the **degree Celsius** (°C). Scientists sometimes make other measurements, including speed, weight, and time.

A. Decide which letter best matches the description. Write *L* for measurements of length, *M* for measurements of mass, *V* for measurements of volume, or *T* for measurements of temperature.

_____ 1. the amount of water in a beaker

_____ 2. the distance from Mars to the Sun

_____ 3. the measure of body heat

_____ 4. the amount of matter in a rock

_____ 5. the height of a moss plant

_____ 6. the amount of rain in a storm

_____ 7. the coldness of dry ice

_____ 8. how far an object falls to the ground

_____ 9. the height of a mountain

_____ 10. the amount of matter in a chunk of carbon

Scientists use scales to make measurements. A **scale** is a series of lines marked on a measuring tool in a regular, ordered way. Different scales are used for different units of measure. Most scales are carefully made so that they allow you to make precise measurements.

B. Figure 2-2 shows laboratory equipment with four different scales. Refer to it as you respond to Items 11 to 17.

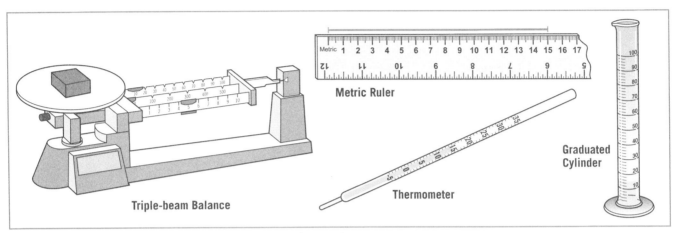

▲ **Figure 2-2** Many pieces of laboratory equipment contain a scale.

11. What do each of these scales measure—length, volume, mass, or temperature?

    a. _____ graduated cylinder   c. _____ ruler

    b. _____ triple-beam balance   d. _____ thermometer

12. The four scales measure different things. In what ways are they alike?

    _____

13. What number do you read on all four scales? _____

14. Make a mark on each scale where it would read 10.

15. On the thermometer, each mark measures one unit. How many marks are there between units of measure on each of the other three scales?

    a. _____ graduated cylinder   c. _____ ruler

    b. _____ triple-beam balance

16. Study each scale closely. Then, make a mark where you think each would read 16.5.

17. Think of the scales you have seen used at stores, at home, or at school. List and

    describe four of them. _____

    _____

Lesson 1 Measuring   41

Practice Measuring

To answer the questions on this page, it would be best to use a metric ruler. If you do not have one, use the drawing on the edge of page 56.

The scale on a metric ruler is marked in centimeters (cm) and millimeters (mm). A **centimeter** is equal to 1/100 of a meter. The ruler on page 56 is 15 cm long. Each centimeter on the ruler is divided into ten smaller units. Each of these smaller units is a millimeter. A **millimeter** is equal to 1/10 of a centimeter, or 1/1,000 of a meter.

A measurement of 6 cm plus 3 mm would be written as 6.3 cm. So, a line that is 20 cm plus 9 mm long is described as being 20.9 cm long.

A. Use a metric ruler and Figure 2-3 to make the following measurements. Answer the questions in the spaces provided.

1. Look at the longer line on the left side of this page. Place your ruler next to it, with the bottom end of the line exactly at the 1-cm mark. What mark on the ruler touches the top end of the line?

   _____

2. What is the length of that line? _____

3. Place your ruler on the shorter line with the bottom of the line touching the 1-cm mark. What mark does the top of the line touch? _____

4. What is the length of the shorter line? _____

5. When measuring, it is best to use marks within a ruler rather than starting at the end of the ruler. Why do you think that is important?

   _____

▲ Figure 2-3

B. Use your ruler to measure a textbook, your desktop, and your notebook. Answer the questions in the spaces provided.

6. How can you make an accurate measurement with a ruler that is shorter than

   the object you want to measure? _____

   _____

7. Using the method you described in Item 6, what is the measurement of

   the longer side of your desktop? _____

   _____

8. Measure the shorter and longer sides of your notebook. What is the difference?

   longer side = _____ shorter side = _____ difference = _____

Thinking About Measuring

In science, accurate measurements are very important. Measuring tools with precise scales are necessary. However, the way a person uses the tools to take measurements also makes a difference.

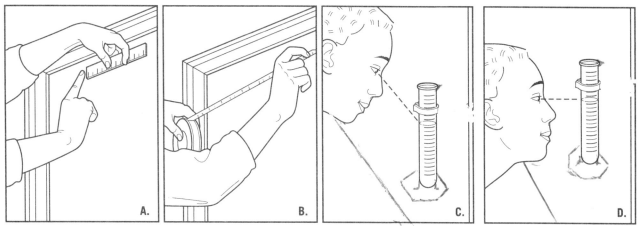

▲ **Figure 2-4** In each pair, one method of measuring is more accurate than the other method.

A. Study each pair of pictures. Then, answer Items 1 and 2.

1. Which way is more accurate, **A** or **B**? _____ Why? _____

   _____

2. Which way is more accurate, **C** or **D**? _____ Why? _____

   _____

When measuring, a scientist must decide which unit of measurement to use. It would not be useful to measure the distance from Mexico City to New York in millimeters.

B. Study the numbered items on the left. Then, draw a line from each item to the unit of measurement that you think would be proper to use. The first one has been done for you.

   3. mass of a beaker   meters (m)

   4. rainfall in a day   milliliters (mL)

   5. liquid in a test tube   grams (g)

   6. altitude of a mountain   centimeters (cm)

Extending Your Experience

Answer the following in complete sentences. Use a separate sheet of paper.

1. An automobile speedometer measures kilometers per hour (kph). What are the two types of measurements used to form the unit of measurement for speed?

2. Check the volume printed on a bottle of juice. Pour the drink from the bottle into a graduated beaker to measure it. Was the printed volume correct?

LESSON 2: Estimating Measurements

Reading for a Purpose
- How can you make a useful estimate?
- Why use references when estimating?

Terms to Know
estimate (EHS-tuh-miht) an educated guess
estimate (EHS-tuh-mayt) to make an educated guess
reference (REHF-uhr-uhns) something you know well that can be used to make a comparison

Using References to Estimate

An **estimate** is an educated guess as to the amount or extent of something. Estimates are made when there is no need to be exact. However, estimates still must be fairly accurate to be useful. When you make such an educated guess, you are said to **estimate**. Estimating is an important skill for gathering information.

Estimates in science usually concern qualities that *could* be measured exactly. Scientists estimate size, weight, height, width, extent, age, and so on. Sometimes an estimate is used in predictions. For example, a volcanologist may estimate how many years it will be before a volcano erupts again.

When you estimate, you should keep a reference in mind. A **reference** is something you know that can be used to make comparisons. Many times the reference you use when estimating is your memory. For example, an easy reference for estimating distance is to think of the length of a meter stick.

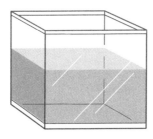

2 liters

1 meter

▲ **Figure 2-5** A meter stick can be a reference for the height of a basketball hoop. A 2-L bottle can be a reference for the volume of a fish tank.

A. Study the pictures in Figure 2-5. Notice that the measurements of two familiar references are given. Compare those references to the other objects.

1. Estimate the height of the basketball hoop, using the meter stick as a reference.

   _____

2. If a meter stick is 1 m, how many meters is the basketball hoop? _____

3. Estimate the volume of water in the fish tank, using the bottle as a reference.

   _____

4. If the bottle holds 2 L, about how many liters will the fish tank hold? _____

44 | Chapter 2 Working As a Scientist

For centuries, parts of the human body have been used as measuring "tools." For example, place your hand on a flat surface and spread your fingers as wide as you can. The distance from the tip of your little finger to the tip of your thumb is called a span. You can also estimate how many foot-lengths it is between two objects.

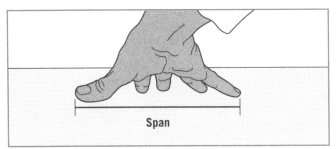 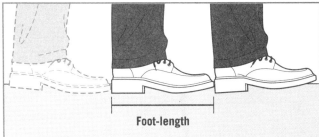

▲ Figure 2-6 A span is a reference using your hand. A foot-length is a reference using your feet.

B. Use Figure 2-6 as a guide. Complete the following sentences.

5. Estimate how many spans wide you think your desk is. The desk is approximately _____ spans wide.

6. Now, check your estimate by performing a span measurement. The desk is exactly _____ spans wide.

7. Measure the width of your span in centimeters. Your span is exactly _____ wide.

8. How many centimeters is the width of the desk? (Multiply the width of your span by the number of spans across the desk.) _____ cm

9. Estimate how many foot-lengths you think it is from your desk to the front of the classroom. _____ foot-lengths

10. What is the actual number of foot-lengths? _____ foot-lengths

11. How many centimeters is your normal foot-length? _____ cm

12. How many centimeters is it from your desk to the front of the classroom? _____ cm

C. Many things can be estimated: time, weight, distance, and so on. Read the following questions carefully. Then, answer them in complete sentences.

13. What reference could you use to estimate 1 minute (60 seconds)? Have a friend time your estimate. How close was your estimate?

    _____

14. What reference could you use to estimate the weight of your science textbook?

    _____

Lesson 2 Estimating Measurements | 45

Practice Estimating

A. Have a friend stretch his or her arms out as wide as possible. Then, answer the following questions.

1. Estimate the distance from your friend's right middle fingertip to the left middle fingertip in centimeters. _____ cm

2. Measure that distance using a tape measure or a meter stick. _____ cm

3. Do you think the distance from fingertip to fingertip is less or more than your friend's height? Measure your friend's height and compare the two distances. _____ cm

4. Using your friend's arm stretch as a reference, estimate the distance of your own arm stretch in centimeters. _____ cm

5. Now, have your friend measure your arm stretch. _____ cm

6. Could knowing the measurement of your arm stretch be useful? What distances could you estimate in arm stretches? _____

▲ Figure 2-7
You can use a container as a reference to estimate the amount of liquid in a bucket.

B. Think of a half-gallon carton of milk as a reference. (A half-gallon container holds almost 2 L.) Then, answer the following questions.

7. Estimate how many liters of water a bucket holds. Compare the half-gallon carton with the bucket you use to wash the family car. Your estimate is _____ L.

8. Use a milk carton to fill a bucket with water. How many cartons does it take to fill the bucket? _____ cartons

9. How many liters does the bucket hold? _____ L

10. What other references would be good for estimating how much liquid something holds? List three. _____

11. Would a milk container be a good reference for estimating the amount of water in a swimming pool? Explain your answer. _____

46 | Chapter 2 Working As a Scientist

Thinking About Estimating

Estimates are made when you need to know about how long something is or about how heavy something is. Sometimes an exact measurement is necessary.

A. Consider the following items. Which could be estimated and which should be measured exactly? Write *E* for estimate or *M* for measurement.

_____ 1. the distance between two cities for making a map

_____ 2. the room temperature in your classroom laboratory

_____ 3. the height of a clamp on a ring stand

_____ 4. the amount of a liquid solution to be used in an experiment

_____ 5. the time needed to complete an experiment

_____ 6. the amount of lava produced by a volcano

B. Answer the following question in complete sentences.

7. If you wanted to know the weight of a rock, would a good reference be a loaf of bread that is about the same size? Explain your answer.

_____

_____

Extending Your Experience

Answer the following in complete sentences. Use a separate sheet of paper.

1. Find a science article in a newspaper, in a magazine, or on the Internet. Make a list of any estimates that were used in the article.

2. Many daily newspapers print hourly temperatures for every day. Estimate the temperature at noon today. Then, check tomorrow's newspaper for the exact temperature at noon.

3. Interview an adult. What estimates are made in his or her job? What, if any, exact measurements are made in the job?

4. Estimate the time needed to walk from your home to a friend's home. Then, check the exact time. How close was your estimate? Would your estimate be the same if 50 people were walking the same route with you?

5. Measure the length of your thumb. List three things you could estimate using the length of your thumb as a reference.

Lesson 2 Estimating Measurements

LESSON 3 — Using the Metric System

Reading for a Purpose
- What do you know about the metric system?
- How do you convert metric measurements within the system?

Terms to Know

unit of measure a standard quantity; a quantity measured by each unit agreed upon by scientists

decimal a number written using the base ten

convert to change from one unit to another

Converting and Comparing Metric Units

In Lesson 1, you learned that scientists make measurements in meters, liters, and grams, as well as other units of measure. A **unit of measure** is a standard quantity. That is, the quantity measured by each unit is agreed upon by scientists.

The metric system of measurement is a **decimal** system. This means that the units of the system are based on 10. The meter (length or distance), the liter (volume), and the gram (mass) are basic units of the metric system. All other units are derived from, or taken from, these basic units.

Look at Figure 2-8. It contains some metric units of length, volume, and mass. Remember the symbol for each unit.

Some Metric Measures		
Length	**Volume**	**Mass**
kilometer (km) (1,000 meters)	kiloliter (kL) (1,000 liters)	kilogram (kg) (1,000 liters)
hectometer (hm) (100 meters)	hectoliter (hL) (100 meters)	hectogram (hg) (100 meters)
dekameter (dkm) (10 meters)	dekaliter (dkL) (10 liters)	dekagram (dkg) (10 liters)
meter (m)	**liter (L)**	**gram (g)**
decimeter (dm) (0.1 meter)	deciliter (dL) (0.1 liter)	decigram (dg) (0.1 gram)
centimeter (cm) (0.01 meter)	centiliter (cL) (0.01 liter)	centigram (cg) (0.01 gram)
millimeter (mm) (0.001 meter)	milliliter (mL) (0.001 liter)	milligram (mg) (0.001 gram)

▲ **Figure 2-8** The metric measures for length, volume, and mass are based on 10.

A. Study Figure 2-8. Then, answer Items 1 to 3 in the spaces provided.

1. A kilometer equals 1,000 m and a kilogram equals 1,000 g. What can you conclude about the meaning of the prefix *kilo-*? _____

2. Define the prefixes.

   a. *deci-* = _____   b. *milli-* = _____   c. *centi-* = _____

48 | Chapter 2 Working As a Scientist

3. What are the symbols for these units?

   a. kilometer _____   c. millimeter _____   e. liter _____

   b. gram _____   d. kilogram _____   f. meter _____

B. For each of the following pairs of units, place a (✓) by the larger unit. The first one has been done for you.

4. a. gram _____   b. kilometer _____   c. milliliter _____

   kilogram __✓__   decimeter _____   deciliter _____

Sometimes you have to change, or **convert**, from one unit of measure to a smaller or larger unit. Converting in the metric system is easy because the metric system is based on tens. There are *always* ten smaller units in the next larger unit. To convert any metric unit to the next smaller unit, you multiply by 10. Look at Figure 2-9.

Conversion Table		
Length	Volume	Mass
1 m = 10 dm	1 L = 10 dL	1 g = 10 dg
1 dm = 10 cm	1 dL = 10 cL	1 dg = 10 cg
1 cm = 10 mm	1 cL = 10 mL	1 cg = 10 mg

▲ **Figure 2-9** Table of metric conversions

What if you had to convert meters to millimeters? Multiply by 10 each time you pass a unit of measure on your way from meters to millimeters.

C. Convert the following, using the table in Figure 2-9.

5. Convert 6 m to millimeters.

   6 m = _____ dm = _____ cm = _____ mm

6. a. 1 L = _____ mL   b. 1 g = _____ mg   c. 1 cm = _____ mm

Sometimes you will need to convert small units to larger units. To do this, divide by 10 each time you pass a unit of measure.

D. Convert each small unit to the larger unit indicated.

7. 5,000 m to kilometers

   5,000 m = _____ dkm = _____ hm = _____ km

8. a. 6,000 L = _____ kL   b. 3,000 mg = _____ g

Practice Using Metric Units

A. Convert each of the following by writing the correct number in the blank. Use a separate sheet of paper to show your work.

1. 6 dg = _____ cg = _____ mg

2. 1,000 m = _____ km

3. 1,000 dL = _____ hL

4. 1,000 cg = _____ dkg

5. 1 m = _____ mm

6. 10 L = _____ cL

7. 1 kg = _____ g

8. 1 g = _____ dg

9. 1 m = _____ cm

10. 1 dg = _____ mg

11. 3,000 mg = _____ g

B. For Items 12-15, place a (✓) by the larger measurement. The first one has been done for you. Use a separate sheet of paper to show your work.

12. 5 kg ✓

    500 g _____

13. 2,000 L _____

    1 kL _____

14. 160 m _____

    16 dm _____

15. 12 cm _____

    60 mm _____

Scientists often need to make comparisons. When comparing measurements, the units must be the same.

C. Use Figure 2-10 to answer Items 16 to 18.

Measurements of Two Metals		
	Aluminum	Iron
Mass	15 dg	6 g
Length	40 cm	50 mm
Thickness	16 mm	8 cm

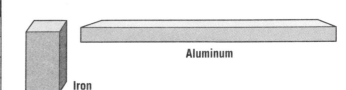

▲ Figure 2-10 To compare measurements, the units must be the same.

16. a. Which sample has more mass? _____

    b. How much more mass does it have? _____

17. a. Which sample is longer? _____

    b. How much longer is it? _____

18. a. Which sample is thicker? _____

    b. How much thicker is it? _____

Thinking About Using Metric Units

The metric system is useful because it is based on the same idea as our number system. Our number system is a decimal system. We count in ones, tens, hundreds, thousands, and so on.

One way of converting metric units is to move the decimal point to the left or to the right. Look again at the chart in Figure 2-8 on page 48. If you go up three larger units, move the decimal point three places to the left. If you go down two smaller units, move the decimal point two places to the right. To convert liters to milliliters, move the decimal point three spaces to the right.

$$1.0000 \text{ L} = 1,000.0 \text{ mL}$$

Use the chart in Figure 2-8 on page 48 to answer Items 1 to 3.

1. To convert a measurement from meters to kilometers, would you move the decimal point to the left or to the right? _____

2. To convert grams to centigrams, would you move the decimal point to the left or to the right? _____

3. Convert these units by moving the decimal point.

    a. 1.6000 km = _____ m   b. 146.8 mg = _____ g   c. 1,638.7 cL = _____ L.

Extending Your Experience

Answer the following in complete sentences. Use a separate sheet of paper.

1. The system of measurement used by most people in the United States is called the English System of Weights and Measures. Converting in the English system is very awkward. There is no logical relationship between the number of smaller units to a larger unit. For example:

    1 mile = 1,760 yards; 1 yard = 3 feet; 1 foot = 12 inches

    Convert 1 mile to inches. Then, find out how to convert 2 gallons to cups.

2. Is it possible to convert liters to kilometers? Explain your answer.

3. Converting from one metric unit to another is often done to make the measurement easy to use. These two measurements are equal:

    0.0067 m = 6.7 mm

    Which would be easier to work with? Explain your answer.

4. The metric system was first adopted by a nation in 1799. Find out in which country it was first used. Who was the leader of that country in 1799?

LESSON 4: Practicing Safety in the Laboratory

Reading for a Purpose
- Why are rules important when working in a laboratory?
- What are some laboratory safety rules?

Term to Know
flammable can catch fire easily and burn fast

Laboratory Safety

Working in a science laboratory should be an interesting and rewarding experience. In the laboratory, you can make things happen. You learn by doing. However, in the laboratory you work with equipment and materials that can be dangerous if not handled properly. In order to be safe in the laboratory, there are certain rules and procedures that must be followed.

SAFETY RULES

A. Read the following list of safety rules carefully. After reading all of them, review each rule. Discuss with the class why each rule is important. Understanding the reasons for a safety rule will make you aware of why it is important to follow that rule.

1) Wear appropriate clothing. Always wear safety goggles and a laboratory coat or apron in the laboratory. Loose garments and jewelry should be removed. Long hair should be tied back or covered.

2) Follow instructions completely. Perform only the laboratory activities assigned by your teacher. If you are in doubt about a procedure, ask your teacher.

3) Keep your work area clean and neat. **Flammable** materials, those that catch fire easily, should be kept away from open flames. Unnecessary items should be stored away from your work area. Clean up spills immediately according to your teacher's instructions.

4) Set up laboratory apparatus carefully. Follow instructions from your teacher or laboratory manual. Make sure ring stands and similar pieces of apparatus are secure.

5) Use proper handling equipment. Tongs, test-tube holders, clamps, and so forth can protect you from burns and spills.

6) Dispose of waste materials properly. Your teacher will explain how to dispose of all materials.

7) Know where emergency equipment is and how to use it. If clothing catches fire, do not run. Smother the fire with a towel or blanket, or drop to the ground and roll. Use lots of water to rinse chemical spills.

8) Clean up your work area at the end of the laboratory period. Turn off water. Disconnect electrical equipment. Wash your hands.

SAFETY RULES (continued)

9) Check the labels on chemical bottles carefully. Be sure you are using the correct chemical. Take only as much as you need.

10) Avoid getting chemicals on your skin and clothing. Hold containers away from your body when transferring the contents. If chemicals do spill on you, wash the area immediately and report the spill to your teacher.

11) When mixing acid and water, never add water to acid. Add acid to water slowly to prevent spattering.

12) Never taste or smell any substance unless instructed to do so by your teacher. Keep your face away from containers holding chemicals. Activities involving poisonous vapors should be performed under an exhaust hood.

13) Handle glass equipment carefully. Carry glass tubing vertically. Use safety equipment to hold and transfer hot glass.

14) Lubricate glass tubing before inserting it into a stopper. Water or glycerin helps the glass tubing to move easily to prevent breakage.

15) Never force stoppers into or out of glassware. Use a gentle, twisting motion to ease stoppers into or out of test tubes, beakers, and so forth.

16) Never heat a stoppered piece of glassware. Substances in the glassware will expand when heated, causing the glass to break or the stopper to fly out.

17) Do not clean up broken glass with your bare hands. Use a brush and dustpan, and dispose of the glass in a safe place.

18) Turn off heat sources when not in use. Never leave heat sources unattended.

19) Always point test tubes away from yourself and others. Never look into a container that is being heated.

20) Allow time for equipment to cool before handling it. Always use an insulated mitt or the proper equipment (tongs, clamps) to handle a heated apparatus.

21) Be sure electrical equipment is disconnected before handling it or adding circuits to it.

22) Make sure the area around all electrical equipment is dry. Water conducts electricity. Getting an electrical shock is very dangerous.

23) Check all electrical cords before using them. Make sure all insulation, plugs, outlets, and wiring are in good order. Report all faulty equipment to your teacher. Do not use any faulty electrical equipment.

24) Always work in the laboratory with teacher supervision.

B. List some safety rules that are not included in the previous list.

_____

_____

_____

Practice Using Safety in the Laboratory

Always follow your teacher's instructions carefully. Many pieces of equipment are easily broken. Some activities involve potentially harmful materials. When working in the laboratory, you must follow certain rules to protect yourself, your classmates, and valuable equipment.

▲ **Figure 2-11** It is a good thing Einstein's lab never really looked like this.

A. Look carefully at Figure 2-11. What is wrong with this picture? In the spaces provided, list the number of the safety rule (or rules) not being followed. Then, explain how to fix the problem.

1. Things wrong _____

_____

2. How to fix _____

_____

B. The following are a few rules. Write the names of at least two different pieces of equipment in the space provided. The rule should apply to the items you name. You can use items shown in the lesson or other items used in science.

3. Wear appropriate clothing when using items that could cause personal injury.

_____

54 | Chapter 2 Working As a Scientist

4. Before using any glassware, check it for damage, such as chips, cracks, or breaks. Tell your teacher right away if anything is damaged. _____

5. Store all pieces of equipment that could be broken easily. Safe storage protects the equipment, yourself, and other people. _____

6. Never use a delicate, complex device unless you know how to operate it. _____

7. Write two more rules for dealing with the safety or care of equipment.

Thinking About Safety in the Laboratory

There are many items in your home that require you to be as safety conscious as you must be in the laboratory. Take the time to go through your home and identify some items with which you and your family need to practice safety. The following suggestions may help you to organize your search.

Next to each topic, list 3 or 4 items from your home that require safety precautions be followed.

1. medicines _____
2. electrical devices _____
3. household cleaners _____
4. glassware _____

Extending Your Experience

Answer the following in complete sentences. Use a separate sheet of paper.

1. Look through your science textbook. How does the publisher indicate that there might be danger when performing an experiment or investigation? What icons, or characters, are used to alert you to the various safety issues that may arise while you work in the laboratory?

2. Read an article from a newspaper, a magazine, or on the Internet that discusses some science topic with possible hazards. What safety precautions are being followed or suggested in order to avoid possible accidents?

LESSON 5 Working in the Laboratory

Reading for a Purpose
- How can you recognize different pieces of apparatus used in science?
- How many rules can you list about using this apparatus safely?

Term to Know
apparatus (ap-uh-RAT-uhs) equipment, such as tools and devices, used to perform a task

Identifying Apparatus

Scientists use many different kinds of apparatus. **Apparatus** are the tools and devices used to perform scientific tasks. With apparatus, scientists test, investigate, measure, analyze, identify, and store the objects they study.

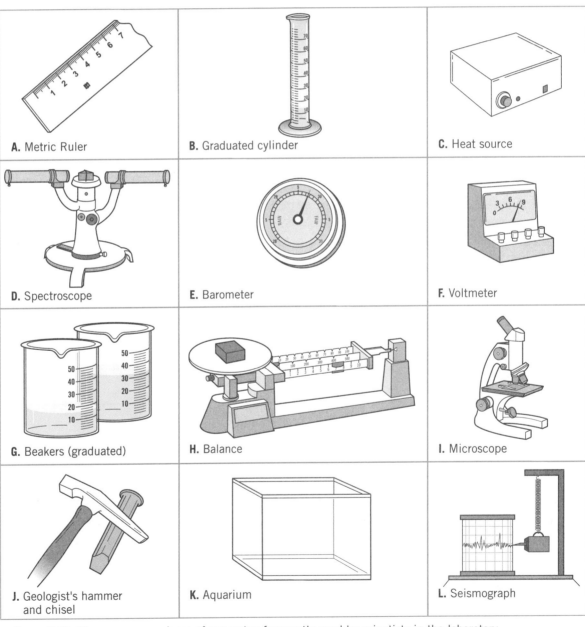

▲ Figure 2-13
A metric ruler

A. Metric Ruler
B. Graduated cylinder
C. Heat source
D. Spectroscope
E. Barometer
F. Voltmeter
G. Beakers (graduated)
H. Balance
I. Microscope
J. Geologist's hammer and chisel
K. Aquarium
L. Seismograph

▲ **Figure 2-12** These are some pieces of apparatus frequently used by scientists in the laboratory or for work in the field.

56 | Chapter 2 Working As a Scientist

Study the pictures in Figure 2-12. In each blank, write the letters of the apparatus that best answer each question. The first one is done for you.

1. Which pieces of apparatus pictured have some kind of scale? __a, b, e, f, g, h, l__
2. Which pieces of apparatus contain glass that could break? _____
3. Which pieces have movable parts? _____
4. Which pieces could be used to study solid elements? _____
5. Which pieces could be used to study light? _____
6. Which pieces could be used to measure? _____
7. Which pieces could be used to study liquid substances? _____
8. Which pieces allow you to see a sample of matter more clearly? _____
9. Which pieces require an electric current in order to operate? _____

Practice Identifying Apparatus

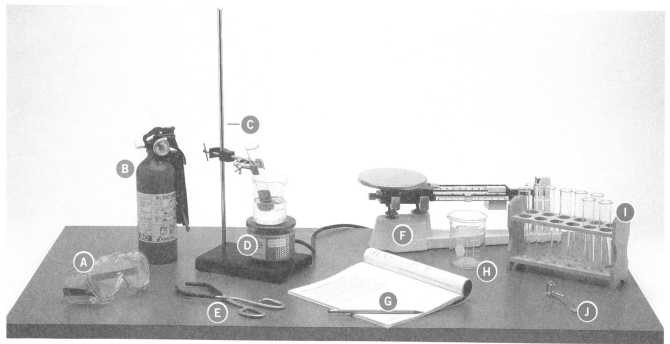

▲ Figure 2-13 You can find different apparatus in a laboratory.

A. In the spaces provided write the letter that matches each piece of apparatus in Figure 2–13.

_____ 1. beaker
_____ 2. heat source
_____ 3. beaker tongs
_____ 4. test tubes in rack
_____ 5. ring stand and clamp
_____ 6. test tube holder
_____ 7. balance
_____ 8. safety goggles
_____ 9. manual and pencil
_____ 10. fire extinguisher

Lesson 5 Working in the Laboratory | 57

B. Look at Figure 2-13. Answer Items 1 to 7 in the spaces provided.

11. Which apparatus can you use to make measurements? _____
_____

12. What materials do you need to record data? _____

13. Which apparatus protects you? _____

14. Which apparatus are useful for holding liquid or powdered substances?
_____
_____

15. Which piece of equipment in the picture is the most dangerous at the moment? Why? _____
_____

16. Would a fan or fire extinguisher be useful during the experiment in Figure 2-13? Explain your answer. _____
_____
_____

17. What other equipment might be useful when performing the experiment in Figure 2-13? _____
_____
_____
_____

Thinking About Complex Apparatus

Test tubes, graduated cylinders, and balances are examples of scientific apparatus you are likely to find and use in your classroom science laboratory. Some types of scientific apparatus are used mostly by professional scientists. These machines are complex, expensive, and provide very special information.

The two devices shown in Figure 2-14 are complex apparatus used in science. A description of how each works is given.

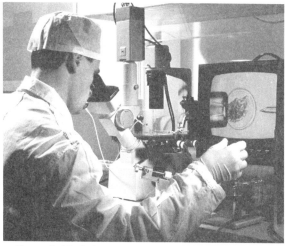 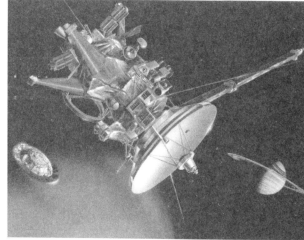

▲ Figure 2-14 Scientists use highly specialized, expensive equipment.

Study the pictures in Figure 2-14 and read the descriptions. Then, answer the following questions.

1. A high-powered microscope can magnify matter thousands of times its actual size. This helps the scientist perform microsurgery on a single cell. How could a high-powered microscope be useful to scientists?

   _____

   _____

2. A space probe can be sent to planets in the solar system to collect information. A space probe can send a robot to test the atmosphere and soil. It also can take pictures. The results are sent back to Earth by radio signals. What can a scientist learn about the planets by using such an apparatus?

   _____

   _____

Extending Your Experience

Answer the following in complete sentences. Use a separate sheet of paper.

1. When was the first thermometer made? Who made it? Use reference materials to answer these and other questions you may have about this useful and common apparatus.

2. Find an article in a magazine or newspaper about a science topic, such as heat energy, medical research, or astronomy. What apparatus is mentioned? What apparatus was used to collect the information reported in the article?

3. What is the difference between regular eyeglasses and safety goggles?

4. Look in science magazines or online for advertisements of science equipment. Make a list of the tools and machines an amateur scientist can buy through the mail or over the Internet.

LESSON 6 Sampling

Reading for a Purpose
- What is the value of sampling?
- How do you estimate a whole amount by studying a sample?

Terms to Know
census (SEHN-suhs) a study of each and every part of a group

sample a study of just enough parts to understand the group

random sampling (RAN-duhm SAM-plihng) every part of the group has an equal chance of being included as part of the sample

systematic sampling the use of a system to take the sample

biased (BY-uhst) **sample** a sample that contains errors favoring one result over another

Information Through Sampling

Scientists can study a group of things in one of two ways. They can take a **census**. This is accomplished by counting each individual part of the whole. They can also take a **sample**, which is a study of just enough parts to understand the group. A sample is useful when it is difficult to take a census.

Scientists have developed many ways to take samples. A basic way is random sampling. In **random sampling**, each object in a group has an equal chance of being studied or counted as part of the sample.

Look at Figure 2-15 showing bees on a honeycomb. It would take too long to count every bee. So the picture has been divided into sections.

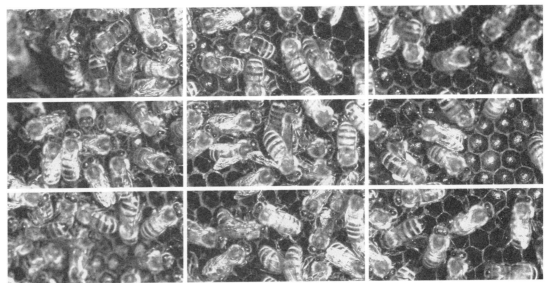

▲ **Figure 2-15** Bees make honey from nectar collected from flowers. The photograph has been divided into sections for sampling.

A. Count the bees in three sections of Figure 2-15. Then, figure out how many bees are probably in the whole picture. To make the sample a random one, close your eyes and put your finger on the picture. Count the bees in the section you touch. Do this for three sections, recording the results in Item 1.

1. How many bees are there in your three samples?

   a. First sample _____   b. Second sample _____   c. Third sample _____

2. Add the three to get the total number in your random sample. _____

3. Find the average number of bees in your three samples.

Sum of Three Samples	Divided by	Number of Samples	Equals	Average
_____	÷	3	=	_____

4. Now, find the total number of bees that are probably in the picture.

Average	Multiplied by	Total Number of Sections	Equals	Estimated Total Number of Bees
_____	×	9	=	_____

5. Compare your estimated total number of bees with a friend's figure.

   What is the difference? _____

Taking samples is one way of estimating. That is why your answer for the total number of bees may be different from someone else's. The more sections counted, the closer the estimate comes to the actual figure. A more common way for scientists to take a sample is called systematic sampling. In **systematic sampling**, some sort of system is used to take the sample.

B. Look at Figure 2-15 again. This time, count every other section. Start with Section 1. Next, count Section 3. Then, count Section 5.

6. How many bees are there in your three samples?

   a. First sample _____   b. Second sample _____   c. Third sample _____

7. Add the three samples to get the total number in your systematic sample. _____

8. Find the average number of bees in these three samples.

Sum of Three Samples	Divided by	Number of Samples	Equals	Average
_____	÷	3	=	_____

9. Now, find the total number of bees that are probably in the picture.

Average	Multiplied by	Total Number of Samples	Equals	Estimated Total Number of Bees
_____	×	9	=	_____

10. Compare the total from your systematic sampling to the total from your random sampling. What is the difference? _____

Practice Sampling

Here is a picture of a tabletop covered with marbles (the larger spheres) and beads (the smaller spheres). You can use different sampling methods to estimate their numbers.

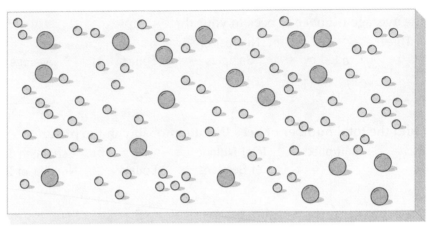

▲ Figure 2-16 Marbles and beads

A. Refer to Figure 2-16 to answer Items 1 to 5.

1. What would be the easiest way to figure out the number of beads in the picture? Why? _____

2. Using a metric ruler, divide Figure 2-16 into 12 equal parts to make a grid system similar to the one used to count the bees. Now, take a random sampling by counting the number of beads in any three sections.

   a. First sample _____ **b.** Second sample _____ **c.** Third sample _____

3. Sum of the three samples _____

4. Now, find the average number of beads in the sample.

Sum of Three Samples	Divided by	Number of Samples	Equals	Average
_____	÷	_____	=	_____

5. What is the estimated total number of beads in the picture?

Average	Multiplied by	Total Number of Sections	Equals	Estimated Number of Beads
_____	×	_____	=	_____

62 | Chapter 2 Working As a Scientist

B. Now, make up a system to take a systematic sampling of five sections of the same grid. Try using four corner squares and one in the middle, or pick five squares that slant across the picture like a staircase. Use a separate sheet of paper.

6. Use the space below to record the number of beads in each sample section. Find the average for the five sections. Then, find the estimated total number of beads.

7. Which sampling do you think was more accurate? Why? _____

Thinking About Sampling

Systematic sampling helps prevent a biased sample. A **biased sample** is one that contains errors that tend to favor one result over another. For example, suppose a scientist wants to know how well all known metals conduct electricity. However, the scientist does not sample gold or silver because these metals are too expensive. That scientist's sampling would be biased because the information about gold and silver would be missing. Inaccurate apparatus can also result in a biased sample. If a thermometer always reads 6°C higher than it should, all the temperature figures in a sample made using that thermometer would be too high.

Answer the following questions in complete sentences.

1. Think of the bee sampling you did earlier. Suppose a person picked three sections to count because those sections did not have very many bees in them. Would that be a biased sample? How would the sample affect the person's answers?

   _____

2. Suppose your pocket calculator was broken, but you did not know it. It always entered 7 when you punched 3. You were using the calculator in sampling the masses of equal volumes of several liquids. Would your sample be biased?

   Explain how the results would be wrong. _____

Extending Your Experience

Answer the following in complete sentences. Use a separate sheet of paper.

1. Look again at Figure 2-16. Find the three sections with the fewest spheres in them and repeat the exercise. Compare that total to your random and systematic probable totals. Which of the three totals is least accurate? Why?

2. Two scientists want to investigate what percent of people have blue eyes. One scientist's sampling includes examining fifty people. The other scientist tests five hundred people. Whose results would you be more willing to believe? Why?

3. An electrical engineer samples the amount of light given off by light bulbs of different wattage. How is the engineer's scientific sampling different from the way a consumer would sample these light bulbs for use in the home?

4. Find out about scientists who watch for pollution levels in the water or air. How do they use sampling in their work?

5. Suppose you wanted to sample the kinds of rocks found around your home. You like red rocks the best. So, you will usually pick up every red rock you see. Think of a system that would prevent your sample from being biased.

LESSON 7 Recording

Reading for a Purpose
- How are record entries made?
- What is the importance of careful record keeping?

Terms to Know
record (REHK-uhrd) a lasting report that keeps information for later use

record (rih-KAWRD) to write down or save information in a permanent way

Keeping Records

Scientists perform experiments to test hypotheses and to pass along the information they discover. What would happen if scientists passed along only the information they remember by simply telling it to another scientist? It is likely the data would be mixed up or some important facts would be left out.

A **record** is a lasting report that keeps information about scientific experiments and methods in good order for later use. To **record** is to write down or save information in any permanent way.

Making an accurate record

- Adds new information to scientific knowledge
- Preserves data for other scientists and students to use
- Lists apparatus and experimental methods, or ways of doing things
- Puts data into an understandable form
- Gives all the information another scientist would need to repeat the experiment

A. Here are two records of classroom temperature taken twice a day for 3 days. Compare the two records. Then, answer Items 1 to 5.

Record A

Classroom Temperatures Week of December 5			
Date	Temperature	Time	Comments
12/5	19°C	10:00 a.m.	Classroom windows face west; thermostat was raised from 14°C (weekend setting) to 20°C at 7:30 a.m.
12/5	22°C	2:00 p.m.	Window shades raised; bright sunlight entering room.
12/6	20°C	10:02 a.m.	One window open 25 cm to ventilate after chemistry experiment.
12/6	20°C	1:59 p.m.	Shades lowered for filmstrip; window now closed.
12/7	18°C	10:00 a.m.	Blower motor in furnace being repaired.
12/7	17°C	2:03 p.m.	Motor still not working; overcast day.

Record B

Temperatures			
Date	Temperature	Time	Comments
12/5	19°C	10	
12/5	22°C	afternoon	
	20°C	10 a.m.	Feels too hot
12/6	20°C	2 p.m.	Feels better
12/7	18°C	midmorning	Something wrong with furnace
12/7	??	2	Forgot to take reading; feels colder

▲ Figure 2-17 Two records of classroom temperatures

1. Which is the better record, Record A or Record B? Why? _____
   _____

2. List some data found in Record A that does not appear in Record B. _____
   _____
   _____

3. Which record would be more useful to another class that wants to compare its room temperature to these readings? Why? _____
   _____
   _____

4. Using only the information given, which record better explains the variations in classroom temperature? _____
   _____

5. List three variables given in Record A that might have affected classroom temperature. _____
   _____

Scientists must be able to trust their own records as well as the records of other scientists. A record that is not accurate and not clear is useless. A good record should be clear both to scientists in other countries and to scientists of the future. Therefore, the information in a record should not include terms that might mean different things to different people. When you write a record, always try to think like one of your classmates. Would that person understand exactly what you mean?

B. Read the following statements. Show those statements which should be included in a record by marking them with a (✓). Show those statements which should not be included in a record by marking them with an *N*.

6. _____ This liquid is hot.

7. _____ The plant in the shade is 4 cm high.

8. _____ The rock is big.

9. _____ A lot of energy was given off.

10. _____ There are 25 cells visible in the sample.

11. _____ The volcano erupted at 4:37 P.M.

12. _____ The white powder weighed about 2.2 g.

13. _____ The fossil was very old.

14. _____ The earthquake measured 5.7 on the Richter scale.

Lesson 7 Recording

Practice Keeping Records

Inez and Trevor work for the National Weather Service. Every hour, they record weather information at their office near Denver, Colorado. Here is a description of a typical morning at the weather office.

Inez and Trevor went on duty at 9:00 A.M. on May 5, 2002. Trevor was sick, but he went to work anyway. At 9:00, Inez noted that the temperature outside was 14°C. Trevor checked the barometer. It read 29.5 in. of mercury. The wind was from the southeast at 15 kilometers per hour (kph). Trevor drank some orange juice. Inez watched a storm move in. At 10:00, it began to rain. The temperature was 17°C and the barometer read 29.3 in. A southeasterly wind was blowing at 20 kph. At 11:00 A.M., Inez and Trevor recorded the barometer reading at 29.1. The temperature was 18°C. A southeasterly wind was blowing at 22 kph. At noon it stopped raining, The temperature was 18°C, the barometer was steady at 29.1 in., and the southeasterly wind had fallen to 10 kph. Exactly 1.1 cm of rain had fallen. Inez and Trevor were hungry. They were glad it was time for lunch.

A. Use the chart that follows to make a record of their findings.

OFFICIAL RECORD

Date: _____   Place: _____

Observers: _____

Time	Temperature	Barometer Reading	Wind Speed	Wind Direction

Precipitation: from _____ to _____

Accumulation: _____

▲ **Figure 2-18** Inez and Trevor may have kept a chart like this one.

B. Look at the information you used to complete the record in Figure 2-18. Then, answer the following questions.

1. Give an example of information in the description that you did not use in the record. _____

2. Why did you not use that information? _____

3. What other measurements might Inez and Trevor have taken that they could have used in the record? _____

Thinking About Keeping Records

The pictures on the walls of caves where early humans lived represent a sort of record. Ancient Greek philosophers, such as Aristotle, kept their records on scrolls, which are rolls of paperlike material. Scrolls were large and difficult to store and tended to fall apart. These early "scientists" did the best they could with the materials they had. These days, records that were once kept on 100 scrolls can be stored on a small part of a single computer disk.

A. Think of what you know about computers to answer the following questions.

1. Scientists who study fossils store their fossil finds in boxes and vaults. How might a computer help them in their work? _____

2. Scientists who study rocketry must determine the amount of force needed to lift a rocket into orbit. How could a computer help them in their work?
_____

B. Computers are not the only apparatus used by modern scientists to make and store records. Can you think of other apparatus scientists might use?

3. List two examples of apparatus used to make records. _____
_____

4. List two examples of apparatus used to store records. _____
_____

Extending Your Experience

Answer the following in complete sentences. Use a separate sheet of paper.

1. Make a record of an experiment or activity you have done in your science class. Be sure to make it complete and clear, so another person could perform the activity and compare results.

2. Paper has long been used to make and keep records. Find out where, when, and how paper was first made. Compare this early method to the way much of our paper is made today.

3. In the role of an astronomer at a university, write a paragraph of instructions for new scientists. Tell them how they should go about making and keeping records of their observations. In general terms, describe what they should put in their records.

4. When scientists observe and measure, they record much information that does not seem important. Why do you think they do that? Might such information ever be useful? Explain your answers.

5. Find an article related to endangered species in a newspaper, a magazine, or the Internet. Make a record from the article. Use only information that should go into a record. Is your record a complete one, or did the article leave out some important information?

Chapter 2 Review

Concept Review

A. Write *E* for something that could be estimated or *M* for something that should be measured.

_____ 1. whether or not there is enough light by which to read

_____ 2. the amount of bacteria in a sample of food

_____ 3. the energy released by an earthquake

_____ 4. the size of a cloud

_____ 5. the temperature at which mercury becomes a solid

B. Study the apparatus shown in Figure 2-19. Then, answer Items 6 and 7.

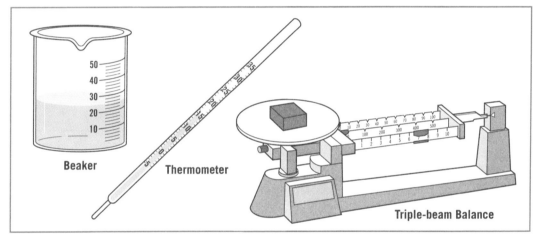

▲ **Figure 2-19** Laboratory apparatus

6. What are each of these apparatus used for?

   a. beaker _____

   b. thermometer _____

   c. triple-beam balance _____

7. What units of measure do each of these apparatus show?

   a. beaker _____

   b. thermometer _____

   c. triple-beam balance _____

C. Answer each of the following questions in complete sentences.

8. You have managed to collect a variety of beetles from at least three different families. How would you do a systematic sampling of your collection to know how many beetles of the Cerambycidae family you have? Explain each step of the process.

_____

_____

_____

9. List two things you should always keep in mind when making a record.

_____

_____

Vocabulary Review

Complete the crossword puzzle using the Terms to Know from this chapter.

ACROSS
1. the basic metric unit of length
4. a measure of distance
5. the measure of how hot an object is
9. something used for comparison when estimating
12. a good guess about something
13. a study of a part of the whole
14. the basic metric unit of mass

DOWN
2. the basic unit of temperature
3. a lasting account
6. tools and machines used by scientists
7. to determine the exact dimensions of something
8. liquid capacity
10. the quantity of matter
11. the basic metric unit of volume

CHAPTER 3

Thinking As a Scientist

Scientists seek to understand natural objects and the cause of natural events. In their investigations, scientists follow certain methods of inquiry. Besides planning and working in a certain way, they must think in a certain way.

Comparing to see how things are alike and different is one way scientists gather information. With this information, scientists can classify things by grouping together those that are similar.

Recognizing patterns is another important skill in science because patterns provide the order necessary for understanding. As scientists gather information, they can begin to make generalizations and draw conclusions. In these ways, scientists can help us all to better understand our world.

In this chapter, you will learn some of the skills that scientists use to add to our knowledge of the world we live in.

SKILLS
- **Comparing** .... LESSON 1
- **Classifying** .... LESSON 2
- **Using Chemical Shorthand** .... LESSON 3
- **Using Guides and Keys** .... LESSON 4
- **Recognizing Patterns in Science** .... LESSON 5
- **Understanding Cause and Effect** .... LESSON 6
- **Concluding** .... LESSON 7
- **Generalizing** .... LESSON 8

▲ **Figure 3-1** Pattern created by DNA test.

SCIENTISTS MUST always think ahead. Scientists like the one in the photograph have thought ahead and introduced the Texas cougar to a population of Florida panthers. The population of panthers was decreasing so severely that the gene pool was no longer safe. Introducing a similar breed may help save the entire population. Genetic data, like that in the inset, shows patterns of breeding in panthers. Veterinarians tranquilize the panthers in order to take blood samples.

LESSON 1 Comparing

Reading for a Purpose
- How can you recognize that objects are similar or different?
- How can you use references to make comparisons?

Terms to Know
comparison an examination of similarities and differences
similar the same
different not alike
reference something you know
standard a value agreed upon by everyone

Making Comparisons

When it snows, all the snowflakes seem to look the same. Yet, if you look closely, you will see great differences among them. Many things look the same at first sight. However, scientists are trained to see how things are similar or different. Exploring the world around them involves making endless comparisons. A **comparison** is an examination of similarities and differences.

A. Study the sets of objects in Figures 3-3 and 3-4. Then, answer Items 1 and 2 in the spaces provided.

1. Compare the two leaves in Figure 3-3.

   a. In what ways are they similar?

   _____

   _____

   b. How are they different?

   _____

   _____

▲ Figure 3-2 Snowflakes share many similarities, but also have many differences.

▲ Figure 3-3 You can see the similarities and differences in these leaves.

2. Compare the two states of matter in Figure 3-4.

   a. In what ways are they similar?

   _____

   _____

   b. How are they different?

   _____

   _____

▲ Figure 3-4 Water can be found in more than one form, or physical state.

Chapter 3 Thinking As a Scientist

A **reference** is something you know. Think about how you used references in making estimates. You can also use references when describing something unknown or unfamiliar. In your description, you can compare the unfamiliar thing to something you already know.

Buckminster Fuller, an inventor and engineer, used geodesic domes in some of his architectural designs. In Figure 3-5, notice the resemblance between this design and the shape of the fullerene model shown at the right. The carbon model was named after Fuller because of the similar shapes of the dome and model. Notice, however, that the size of the molecule is much smaller than any of Fuller's domes. Fullerene is one of the most interesting forms of carbon.

B. Fullerenes can be described by using references. Study the geodesic dome and fullerene in Figure 3-5. Then, read the following sentences. Circle the reference used to describe fullerenes in each sentence.

3. A fullerene resembles a soccer ball.
4. A fullerene is made up of a large number of interconnected carbon atoms.
5. A fullerene has a geometric pattern.
6. The pattern of a fullerene is a series of hexagons.
7. A fullerene has a very stable and strong structure.

References can be used to show other qualities, such as speed, temperature, or size. For example, a snowflake has more sides than a square.

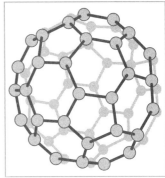

▲ **Figure 3-5** A geodesic dome (top) resembles a fullerene (bottom).

C. Answer Item 8 in the space provided. Then, answer Items 9 to 11 by making comparisons to the references.

8. What can you think of that has more sides than a square? The square is a reference in this example. _____

Reference	Comparison
9. a ten-story building	List two objects that are higher. a. _____ b. _____
10. the mass of a car	List two objects with more mass. a. _____ b. _____
11. a jackhammer operating	List two items that are louder. a. _____ b. _____

Practice Making Comparisons

Look at the two photos in Figure 3-6. The boats are moored in the same location in Canada in both photos. However, there are also some marked differences.

Compare the two scenes in Figure 3-6. Then, answer Items 1 to 4.

▲ **Figure 3-6** Even though these boats are moored at the same location, there are many differences between the two pictures.

1. a. How are the two scenes similar? _____

   _____

   b. How do they differ? _____

   _____

2. From your comparison of the two pictures, what do you think is responsible for the differences in the two scenes? _____

   _____

3. How would your comparison of the two pictures help you identify other differences? _____

   _____

4. What differences would you expect to find in these scenes in December and July?

   _____

   _____

74 | Chapter 3   Thinking As a Scientist

Thinking About Making Comparisons

Every time you measure, you make a comparison. Measurements are made by referring to standards. A **standard** is a value agreed upon by everyone. When you measure the length of a book to be 20 cm, you are comparing that length with the standard known as a centimeter.

A. Place a check (✓) by any of Items 1 to 10 that are standards of measurement.

_____ 1. a liter   _____ 5. a meter   _____ 8. a country mile

_____ 2. a minute   _____ 6. a kilogram   _____ 9. a light year

_____ 3. a whisker   _____ 7. a hand's width   _____ 10. a degree Celsius

_____ 4. a city block

Some comparisons are better than others. A reference that is meaningful to one person might not be meaningful to another person.

B. Answer Items 11 and 12 in the spaces provided.

11. Suppose you were talking to someone who had never traveled outside his or her hometown. Would a good reference for height be the Empire State Building, located in New York City? Explain your answer.

    _____

    _____

12. While discussing air temperature on January first with someone from Brazil, would a good reference be the frigid conditions in Chicago? Explain your answer.

    _____

    _____

Extending Your Experience

Answer the following in complete sentences. Use a separate sheet of paper.

1. In order to describe an elephant to a person without sight, what comparisons could you make? What references could you use?

2. Investigate how standards of measurement were set. Who established the exact length of a meter? How was that standard set? What would happen to laboratory investigations if standard measurements were not used?

3. Write the name of an unusual object at the top of an index card. Then, write a description of the object, using familiar references. Read your description to the class and have them try to guess what the object is.

4. Find a science article in a newspaper, a magazine, or on the Internet. What comparisons are made to help the reader understand the article? What references are used?

LESSON 2 Classifying

Reading for a Purpose
- How can you classify using circle diagrams?
- How do you show relationships among items classified into groups?

Terms to Know

classify (KLAS-uh-fy) to arrange objects into groups based on similarities, or things in common

classifications (klas-uh-fih-KAY-shuhns) groups or classes of objects with certain common characteristics

subgroup small group within a larger group

Classifying Natural Objects

Items can be **classified**, or grouped, in a variety of ways. They are classified by characteristics they all share. For example, glass, rubber, and wood can be classified as insulators. They are all poor conductors of heat and electricity.

Circle diagrams may be used to show how items are related to each other. The diagram in Figure 3-7 groups some objects that reflect light and some objects that absorb all colors of light. The shaded area shows things that absorb some colors of light and reflect some colors of light.

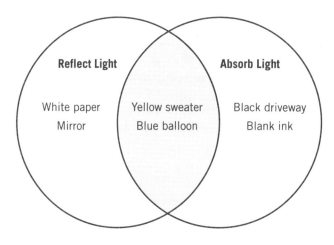

◀ **Figure 3-7** This type of circle diagram, also called a Venn diagram, compares the ways some objects reflect or absorb colors of light.

A. Study the diagram in Figure 3-7. Then, answer Items 1 to 4 in the spaces provided.

1. What things reflect all colors of light? _____

2. What things absorb all colors of light? _____

3. What things both absorb and reflect certain colors of light? _____

4. Where do these objects belong in the circle diagram shown in Figure 3-7? Write the letter of each object in the correct part of the diagram.

   a. green automobile   **c.** purple grapes   **e.** white shirt
   b. pink eraser   **d.** orange crayon   **f.** black slacks

76 | Chapter 3 Thinking As a Scientist

Large groups of items often can be classified into smaller groups called **subgroups**. This can be done by drawing circles inside other circles, as shown in Figure 3-8. In this example, the solar system is the large group. The planets are part of the solar system. Earth is one planet. So, "planets" is a subgroup of "solar system," and "Earth" is a subgroup of "planets."

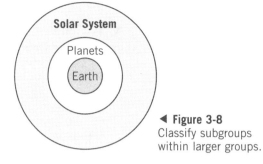

◀ **Figure 3-8** Classify subgroups within larger groups.

B. Use the diagrams in Figure 3-9 to classify the terms listed in Items 5 to 7. Write the names where they belong in the diagrams.

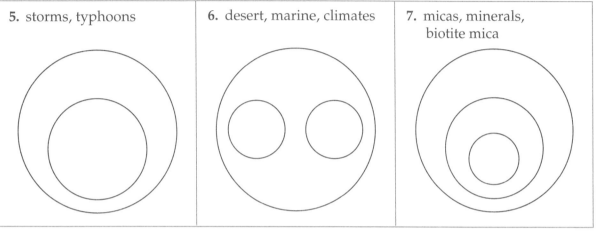

▲ **Figure 3-9** These are circle diagrams used to show classifications of groups and subgroups.

C. Classify the following statement using the circle diagram in Figure 3-10. Then answer Items 8 to 10.

There are many active volcanoes in the world. There are several volcanoes in North America. Some North American volcanoes are active.

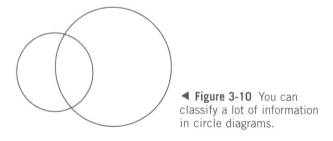

◀ **Figure 3-10** You can classify a lot of information in circle diagrams.

8. What would you label the large circle? _____

9. What would you label the small circle? _____

10. Shade the area that you think would contain the active volcanoes in North America.

Lesson 2 Classifying | 77

Practice Classifying

A. Which of the diagrams in Figure 3-11 show how each of the groups is related? Answer Items 1 to 6 by writing the correct letter or letters in each blank.

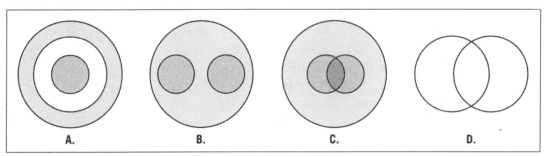

▲ **Figure 3-11** Each of these diagrams represents a classification of information.

_____ 1. amoeba, euglena, protist

_____ 2. sodium, substance, salt, chlorine

_____ 3. elements, gases, halogens

_____ 4. rocks, minerals, quartz

_____ 5. cold air, fog, moisture

_____ 6. solution, mixture, suspension

B. Use the information in the diagram in Figure 3-12 to answer Items 7 to 12. Write *true* or *false* in the space provided.

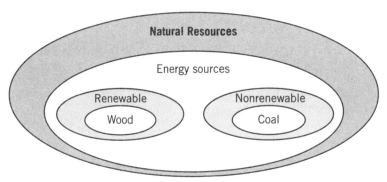

▲ **Figure 3-12** You can classify energy sources as shown.

_____ 7. All energy sources are natural resources.

_____ 8. All energy sources are nonrenewable.

_____ 9. Wood and coal are both energy sources.

_____ 10. Coal is a renewable resource.

_____ 11. Wood is a natural resource.

_____ 12. All natural resources are renewable.

Thinking About Classifying

Almost everything can be classified in more than one way. For example, you could be classified by your grade in school, your hair color, your eye color, your gender, or by the month you were born. For each of the following numbered items, write the letter or letters of way(s) the item could be classified. One has been done for you.

a. size	**d.** composition	**g.** brightness	**j.** depth
b. color	**e.** shape	**h.** speed	**k.** temperature
c. hardness	**f.** location	**i.** age	**l.** luster

_a, e, f, i, j, k_ 1. lakes

_____ 2. stars

_____ 3. minerals

_____ 4. fossils

_____ 5. clouds

_____ 6. storms

_____ 7. volcanoes

_____ 8. caves

_____ 9. bird

_____ 10. light

_____ 11. soil

_____ 12. air masses

_____ 13. coral reefs

_____ 14. alloy

_____ 15. rivers

_____ 16. plant

Extending Your Experience

Answer the following in complete sentences. Use a separate sheet of paper.

1. Use a circle diagram to classify each of the fifty United States by time zone. Write the abbreviations of the states in your circles. Some states include more than one time zone within their borders. Be sure to show that in your diagram.

2. The books in every library are classified by one system or another. Research what system is used in your library. Then, find out how science books are classified in that system.

3. Work with others to identify at least ten science careers. Classify the careers according to a variety of categories, which might include the amount of education required, the courses you need, and the environment where the work is done. Devise some of your own categories. Draw one or more diagrams to show your classifications.

4. Make a picture collection of types of simple machines. You may get your pictures from magazines, the Internet, or you may draw them yourself. Organize your pictures into groups and subgroups. Explain your method of classification.

LESSON 3 Using Chemical Shorthand

Reading for a Purpose
- How can you identify chemical substances by their chemical formulas?
- How can you use chemical equations to identify and name the reactants and products in a chemical reaction?

Terms to Know

element a substance that cannot be broken down into simpler substances through chemical change

chemical symbol the shortened way of writing the name of an element

compound a substance made up of two or more elements that are chemically combined

chemical formula a shorthand way to write the name of a compound using chemical symbols

subscript the number written to the lower right of a chemical symbol in a chemical formula

chemical equation a shorthand way of describing how substances behave in a chemical reaction

coefficient the number that shows how many molecules of a substance are involved in a chemical reaction

balanced equation an equation in which the number of atoms of each element on the left side of an equation is the same as the number of atoms of each element on the right side of the equation

Using Chemical Symbols, Formulas, and Equations

An **element** is a substance that cannot be broken down into simpler substances through chemical change. Ninety-two elements occur naturally. Others have been produced in the laboratory. Alone or in different combinations, elements form all matter.

Scientists use a kind of shorthand to write the names of elements. These shortened names are called **chemical symbols**. Many times, the chemical symbol for an element is the first letter of the element's name. For example, the symbol for oxygen is O. Other times, the chemical symbol is made up of two letters from the name of the element. The symbol for zinc is Zn. Still other times, the chemical symbol for an element comes from the element's name in Latin, or some other language. Fe, the symbol for iron, is from the Latin word for iron, *ferrum*.

A. Use a reference source to answer Items 1 and 2.

1. Find the chemical symbols of these elements. Write the proper symbol for each element in the space provided.

    _____ a. hydrogen   _____ b. magnesium   _____ c. iodine   _____ d. bromine

    _____ e. fluorine   _____ f. copper   _____ g. lithium   _____ h. chlorine

2. Find the names of the elements having these symbols. Write the correct name in the space beside each symbol.

    a. S _____   b. C _____   c. K _____

    d. Al _____   e. Be _____   f. P _____

    g. Au _____   h. He _____   i. Ag _____

80 | Chapter 3 Thinking As a Scientist

Most substances exist in nature as compounds. A **compound** is a substance made up of two or more elements that are chemically combined. A **chemical formula** is a shorthand way to show the elements that make up a compound. In a chemical formula, symbols are used to show the different elements in a compound and the relationships among those elements. Figure 3-13 shows three different compounds. Their names and formulas are given below the drawings.

The formula for each compound tells the kinds of atoms that make up the compound. It also tells the number of each kind of atom. Look at the formula for carbon dioxide in Figure 3-13. The small 2 in the formula shows that there are two atoms of oxygen in this molecule. This number is called a **subscript**. Subscripts are not used to show single atoms. So, a molecule of carbon dioxide contains one carbon atom and two oxygen atoms.

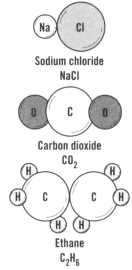

▲ **Figure 3-13** These drawings show three different compounds and their names.

B. Read the directions and answer Items 3 and 4 in the spaces provided.

3. For the following compounds, tell how many atoms of each element are represented in the chemical formula. The first one has been done for you.

   a. H_2O _____two atoms of hydrogen; one atom of oxygen_____

   b. C_2H_6 _____

   c. NaCl _____

   d. $CaCO_3$ _____

4. Write the chemical formulas for the compounds made up of the atoms listed.

   a. one atom of nitrogen, two atoms of oxygen (nitrogen dioxide) _____

   b. one atom of tin, two atoms of fluorine (tin fluoride) _____

   c. one atom of sulfur, three atoms of oxygen (sulfur trioxide) _____

Scientists use symbols and formulas to write chemical equations. A **chemical equation** is a shorthand way of describing how substances behave in a chemical reaction. Look at the following chemical equation:

$$2\ H_2 + O_2 \rightarrow 2\ H_2O$$

This equation tells us that two molecules of hydrogen react with one molecule of oxygen to produce two molecules of water, H_2O. The numbers in front of the hydrogen and water molecules are called coefficients. A coefficient is a number that tells how many molecules of a substance are involved in a chemical reaction. A coefficient of 1 is never written; it is understood to be 1. Thus, there is one oxygen molecule to the left of the arrow in this equation. The arrow in a chemical equation is read as *yields* or *produces*.

C. Study the following chemical equation. Then, answer Items 5 to 7 in the spaces provided.

$$Zn + 2HCl \rightarrow ZnCl_2 + H_2$$

5. How many molecules of zinc (Zn) react? _____

6. How many molecules of hydrochloric acid (HCl) react? _____

7. How many molecules of hydrogen (H_2) are produced? _____

Practice Using Chemical Shorthand

In order for a chemical equation to be correct, it must be balanced. A **balanced equation** is one in which the number of atoms of each element on the left side of the equation is the same as the number of atoms of that element on the right side of the equation. Look at this equation.

$$H_2 + O_2 \rightarrow H_2O$$

Is this equation balanced? To find out, check the number of atoms of each element on both sides of the arrow.

Hydrogen: two atoms on the left, two atoms on the right
Oxygen: two atoms on the left, one atom on the right

The hydrogen atoms are balanced, but the oxygen atoms are not. Therefore, the equation is not balanced. How can we balance it?

The only way to balance a chemical equation is to change the coefficients. NEVER change a subscript. So, to balance the oxygen atoms in this equation, place a 2 in front of the water molecule.

$$H_2 + O_2 \rightarrow 2\ H_2O$$

Now, the oxygen atoms are balanced. There are two on the left and two on the right. But, adding the coefficient has unbalanced the hydrogen atoms. There are now 2 hydrogen atoms on the left and 4 hydrogen atoms on the right. To balance the hydrogen atoms, add the coefficient 2 before the hydrogen molecule on the left side.

$$2\ H_2 + O_2 \rightarrow 2\ H_2O$$

Then, check to see if the equation is balanced.
Hydrogen: four atoms on the left, four atoms on the right
Oxygen: two atoms on the left, two atoms on the right

The equation is now properly balanced.

In the space beside each chemical equation, write *B* if the equation is balanced. Write *N* if the equation is not balanced. Then, balance it.

1. **a.** $Na + Cl_2 \rightarrow 2NaCl$ _____ **b.** $Al + Cl_2 \rightarrow AlCl_3$ _____

c. $H_2 + Br_2 \rightarrow 2\ HBr$ _____   d. $Cu + O_2 \rightarrow Cu_2O$ _____

e. $Si + O_2 \rightarrow SiO_2$ _____   f. $C + H_2 \rightarrow C_2H_2$ _____

2. Check each of the equations to see if it is balanced. Where necessary, write the proper coefficients in the spaces to balance the equation. The first one is done for you.

a. __2__ $Cu +$ _____ $S \rightarrow$ _____ Cu_2S b. _____ $C +$ _____ $Cl_2 \rightarrow$ _____ CCl_4

c. _____ $Fe +$ _____ $Cl_2 \rightarrow$ _____ $FeCl_2$ d. _____ $Mg +$ _____ $N_2 \rightarrow$ _____ Mg_3N_2

e. _____ $Cu +$ _____ $O_2 \rightarrow$ _____ CuO f. _____ $C +$ _____ $O_2 \rightarrow$ _____ CO_2

Thinking About Chemical Shorthand

Answer each of the following in the spaces provided.

1. For each of these compounds, write the name and number of atoms of each element present.

    a. $CaCl_2$ _____

    b. H_2SO_4 _____

    c. $K_2Cr_2O_7$ _____

    d. $NaHCO_3$ _____

2. Balance each of these equations by writing the proper coefficients in the spaces provided.

a. _____ $Fe +$ _____ $O_2 \rightarrow$ _____ Fe_2O_3 b. _____ $Al +$ _____ $Cl_2 \rightarrow$ _____ $AlCl_3$

c. _____ $K +$ _____ $I_2 \rightarrow$ _____ KI d. _____ $N_2 +$ _____ $O_2 \rightarrow$ _____ N_2O_5

Extending Your Experience

Answer the following in complete sentences. Use a separate sheet of paper.

1. Look at the labels of household chemicals. For each ingredient listed as a formula, write the name of the chemical.

2. Use a reference to learn where the names of the elements originated. Report on at least ten different elements.

3. Using references on oceanography, write a report on the elements dissolved in seawater. Include information on what salts would probably form if all the ocean water evaporated.

4. Use litmus paper to test household chemicals for an acid or base. Make a list of chemicals and note how they tested. **Use caution when handling any chemicals.**

5. Look up the law of conservation of matter in a chemistry textbook. Use this law to explain why chemical equations must be balanced.

LESSON 4 Using Guides and Keys

Reading for a Purpose
• How can you use guides to identify natural objects?

Terms to Know
crystal a solid with a definite shape
guide a source that can help direct a person's thinking
key a table that will help you decode or interpret information

Using Keys to Identify Natural Objects

Identifying natural objects is important to a scientist. If an object is unknown, the first task of the scientist is to find out what it is. After the object is named, it can be classified, studied, and compared.

Scientists identify an object by observing its properties and characteristics. Each observed property or characteristic narrows the object for identification. In identifying natural objects, the scientist plays detective. The best scientists are often good detectives.

To identify rocks, a scientist identifies the minerals that make up the rock. Color and crystal shape are clues to the identification of a mineral. A **crystal** is a solid with a definite shape. A mineral's chemical makeup determines its crystal shape. Each mineral has a characteristic crystal shape.

Scientists often use guides to help in identification. A **guide** is a source that can help direct a person's thinking. Within a guide, you may find a **key** that will help you decide or interpret information.

A. Study the mineral crystals in Figure 3-14. Then, use the key that follows to identify the minerals shown. Identify the crystals one at a time. For each crystal, start with Step 1 and proceed as directed. If you read closely, you will identify the crystals correctly. As you identify each crystal, write its name beneath its picture.

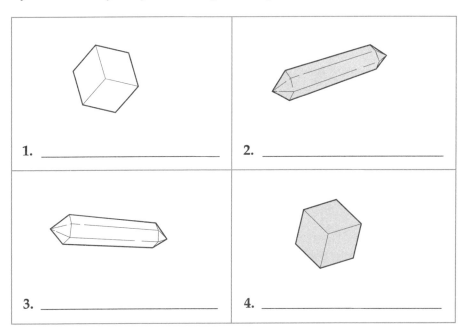

Figure 3-14 ▶
Crystals are solids with definite shapes.

84 | Chapter 3 Thinking As a Scientist

To use the key:
- Look at the first mineral pictured in Figure 3-14.
- Go to Step 1 of the key.
- If the mineral is shaped like a cube, go to Step 2. If not, read the next choice.
- Keep following the instructions until you come to the name of the mineral that fits the description.

Key to Crystals Pictured

Step 1 If the crystal is shaped like a cube .... go to Step 2
   If the crystal is not shaped like a cube .... go to Step 5

Step 2 If the crystal is light .... go to Step 3
   If the crystal is dark .... go to Step 4

Step 3 The crystal is halite.

Step 4 The crystal is magnetite.

Step 5 If the crystal is light .... go to Step 6
   If the crystal is dark .... go to Step 7

Step 6 The crystal is quartz.

Step 7 The crystal is apatite.

B. Below is a diagram of the key you used above. Follow the pathways and write the correct names in the blank boxes.

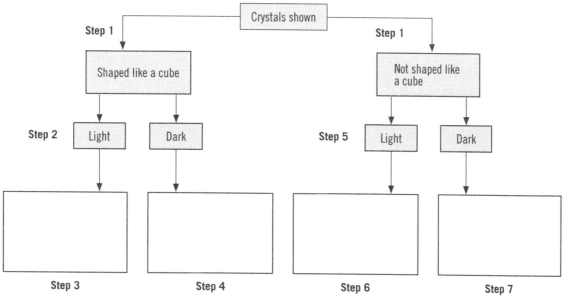

▲ **Figure 3-15** This is a diagram of the key to crystals.

Practice Using Keys

A. Use the following key to identify these birds. Write the scientific name beneath each animal.

▲ **Figure 3-16** You can use a key to identify birds.

1. _____   2. _____   3. _____   4. _____

5. _____   6. _____   7. _____   8. _____

Key to Birds Pictured

Step 1 Can fly .... go to Step 2
           Cannot fly .... name: *Sphenisciformes*

Step 2 Nests in trees .... go to Step 3
           Nests near water................................... name: *Pelecaniformes*

Step 3 Has sharp pointed beak .... name: *Piciformes*
           Has hooked beak .... name: *Strigiformes*

B. Here is a diagram of the key above. Follow the pathways to find the common names of the animals pictured. Write the common names beneath the scientific names in Section A, Items 5 to 8.

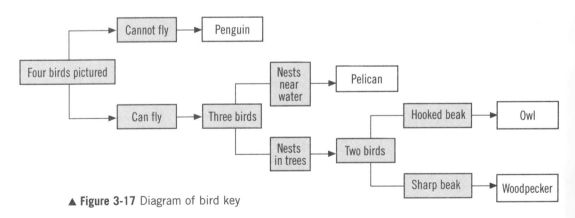

▲ **Figure 3-17** Diagram of bird key

Thinking About Using Keys and Classifying

Keys can be especially helpful to the amateur scientist. Most keys are more complex than those you have used here. You may have to use a combination of pictures, descriptions, and diagrams. However, the general idea is the same. By observing properties and characteristics, you can identify the object in question.

Try making your own key for the following substances. Think of properties and characteristics you can use to identify the substances. Think of the properties and characteristics different substances may have in common.

Substances
Salt, Sulfur, Mercury, Sugar

Step 1 Substance is an element .... go to Step 2
Substance dissolves in water .... go to Step 3

Step 2 _____ .... go to Step 4

_____ .... go to Step 5

Step 3 _____ .... go to Step 6

_____ .... go to Step 7

Step 4 Substance is _____

Step 5 Substance is _____

Step 6 Substance is _____

Step 7 Substance is _____

Write the instructions in the spaces provided. Some of it has been done for you.

Extending Your Experience

Answer the following in complete sentences. Use a separate sheet of paper.

1. Find out about the life and work of Carolinus Linnaeus. You might begin by using an encyclopedia.

2. Using references on oceanography, write a report on the elements dissolved in seawater. Include information on what salts would probably form if all the ocean water evaporated.

3. Find field guides for fossils, shells, planets, reptiles, birds, or other things. Report on a variety of field guides, how they are organized, and how they can be used to aid identification.

4. Write a guide that could help you to identify the various types of clouds. It will have to be longer than the guides you have used in this lesson. Start with the most obvious characteristic of clouds—shape. Then, separate the shapes by altitude. Check in a reference book for the correct information.

LESSON 5   Recognizing Patterns in Science

Reading for a Purpose
- How do you find patterns in science contexts?
- How can you use patterns to make predictions?

Terms to Know

pattern the repeated occurrence of some item

periodic table a table of elements

periodic repeated at regular intervals

Finding and Using Patterns

Scientists search for order in the universe. They observe, name, and classify. They also search for patterns. A **pattern**, in this sense, is the repeated occurrence of some item. Scientists look for patterns to help them make comparisons, see relationships, notice changes, and make predictions. Patterns provide a better understanding of the world.

Early scientists found that elements were the basic substances of all matter. Because the elements could either stand alone or combine in different ways, early scientists looked for ways to organize the elements. These scientists looked for patterns among the elements. In 1869, Dmitri Mendeleev, a Russian scientist, published a table of elements. This table became known as the **periodic table**. **Periodic** means "repeated at regular intervals."

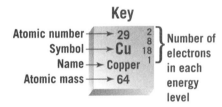

Figure 3-18 ▶ This is a portion of the periodic table, an important use of patterns in science. The key tells you how to read the periodic table.

A. Study the section of the periodic table in Figure 3-18 to find patterns and answer Items 1 to 3.

1. What happens to the atomic numbers of the elements as you read the periodic table from left to right? _____

88 | Chapter 3   Thinking As a Scientist

2. As you read the columns from top to bottom, does the number of electron energy levels increase or decrease? _____ Describe the pattern in the change of the number of energy levels as you read the columns from top to bottom. _____

3. a. In this portion of the periodic table, which element has the lowest atomic mass number? _____

   b. Which element has the highest? _____

   c. What patterns do you observe in the atomic mass numbers of the elements in this portion of the periodic table?
   _____

In the periodic table, elements are grouped in vertical columns. These groups are called families. Each family contains elements with similar properties. The noble gases are one family of elements from the periodic table.

Noble Gases					
Element	Atomic Number	Mass Number	Valence Electrons	Melting Point (°C)	Boiling Point (°C)
helium	2	4	2	−272	−269
neon	10	20	8	−249	−246
argon	18	40	8	−189	−186
krypton	36	84	8	−156.6	−153
xenon	54	131	8	−112	−107
radon	86	222	8	−71	−62

▲ Figure 3-19 This chart shows information about the noble gases.

B. The elements in the noble gas family are arranged from top to bottom in the chart in the same order as they appear in the periodic table. Use the chart to answer Items 4 to 6.

   4. a. How do the atomic numbers change? _____

      b. How do the mass numbers change? _____

   5. a. How do the melting points and boiling points change in the chart?
      _____

      b. What relationship do the melting points and boiling points have to the mass numbers of these elements? _____

Lesson 5 Recognizing Patterns in Science

6. By studying the chart, try to determine what characteristics were used to group the noble gases. _____

Practice Using Patterns

A. Figure 3-20 shows a bird flying. One frame is missing. Which of the frames at the bottom of the figure belongs in the missing space? Write the correct letter in the space provided.

1. _____

▲ **Figure 3-20** You can find missing pieces to a puzzle by recognizing the pattern of the other pieces.

Suppose you are planning to live in another country for a few years. While there, you would like to plant a garden. You know that the part of the country where you are going to live has hot summers and warm winters. However, you need to know what months have the most rainfall. You manage to get the following table, which covers a 4-year period. The numbers in the table show the amount of rainfall in centimeters.

Rainfall For Four Years												
Year	Jan.	Feb.	Mar.	Apr.	May	June	July	Aug.	Sept.	Oct.	Nov.	Dec.
1	10	8	7	4	6	5	8	6	7	7	9	10
2	9	7	6	5	5	6	7	5	6	7	10	11
3	10	8	7	4	4	4	6	7	7	7	9	10
4	10	8	6	4	5	6	7	6	6	7	10	11

▲ **Figure 3-21** This chart shows a record of rainfall over 4 years in the same location.

B. Study Figure 3-21. Do you see any patterns? Place a check (✓) by the statements you think are true.

_____ 2. April usually has the most rain.

_____ 3. December usually has the most rain.

_____ 4. October might be a good time to plant a garden.

_____ 5. August usually has the least rain.

_____ 6. April usually has the least rain.

Thinking About Patterns

Patterns can be found in almost everything. Observing patterns can help determine relationships among data or events. Observing patterns can also help make comparisons or spot change.

Choose two topics from the following list and describe the pattern of each.

| color | numbers | the night sky |
| time | light | growth of a human |

1. First topic _____

2. Second topic _____

Extending Your Experience

Answer the following in complete sentences. Use a separate sheet of paper.

1. Crystals of salt and the mineral galena are cubes. List other examples of things that are cubic in shape.

2. Describe something that has absolutely no pattern.

3. How can patterns become disrupted? Describe three different parts of your everyday life that occur in patterns. List some ways these patterns could be disrupted.

4. Study Newton's laws of motion. Discuss how patterns may have helped in establishing these laws.

5. Find out about the geyser in Yellowstone National Park called Old Faithful. What is Old Faithful's pattern? Why is it called "faithful"?

6. Ancient people used to mark time by the phases of the Moon. What causes the phases? What pattern do the phases follow? How much time passes between each new Moon phase?

LESSON 6 Understanding Cause and Effect

Reading for a Purpose
- How can you recognize cause-and-effect relationships?

Terms to Know
cause an event or condition that brings about an action or result

effect the action or result of a cause

relationship a natural connection between two objects or events

Finding Cause and Effect

Scientists identify, compare, classify, and search for patterns. Yet, perhaps the question most interesting to scientists is not *what*, but *why*. When you investigate the why of something, you are investigating the cause of an effect. A **cause** is an event or condition that brings about an action or result. Such an action or result is called an **effect**. The two words are often linked in the phrase *cause-and-effect relationship*. A **relationship** is a natural connection between two objects or events.

Look at Figure 3-22. Together, these pictures show a cause-and-effect relationship.

▲ **Figure 3-22** On the left is a satellite photo of a hurricane approaching Florida's coast. The photo on the right shows the destructive result produced by a hurricane.

A. Identify the cause and the effect in the following sentences. Circle the cause and underline the effect.

1. The wave action of oceans results in beach erosion.
2. When a warm air mass meets a cold air mass, rain is often the result.
3. A U-shaped valley is produced by an alpine glacier.
4. Metamorphic rocks are created by pressure and heat.

Often an effect has more than one cause. Scientists try to identify as many causes as they can. Sometimes they are able to identify one cause as the major cause.

For example, in April a river overflowed its banks in Boise, Idaho. What caused the flood? Meteorologists knew that the winter had produced more snow than normal. Unusually warm temperatures in early April had melted much of the snow in the mountains. Runoff from the melted snow had caused the river to swell. Then, rain had fallen on the already swollen river. In addition, earlier in the year the government had released more water than usual through a dam near Boise.

B. Reread the previous paragraph on page 92. Then, answer Items 5 to 7.

5. The effect was a flood. List all the causes of that flood. _____

   _____

   _____

6. Could any of the causes have been prevented, or were they all natural?

   _____

   _____

7. In your opinion, what was probably the major cause of the flood? Explain

   your choice. _____

   _____

Firefighters and fire investigators observe the effects of a fire. Then, they try to find out the cause of the fire.

C. Read the following example. Then, complete Items 8 and 9.

A small fire started in a national forest. It soon spread, destroying many thousands of acres. Something caused the fire. There are many reasons why forest fires occur.

8. **Effect**
   What is the effect in this example? _____

9. **Information or data**
   Fire investigators collect much data about fires. Put a C next to the data that could be related to the cause of the fire. Put an E next to the data that could be related to the effect of the fire. Put an X by the data that probably are not related to the fire.

   _____ a. Lightning may have struck some dry underbrush.

   _____ b. A fast-moving river flows through the forest.

   _____ c. Campers did not put out their campfire completely.

   _____ d. The forest is in the northern part of the country.

   _____ e. Firefighters worked for days to put out the fire.

   _____ f. Houses in the surrounding area were threatened by the fire.

   _____ g. Gasoline had been spilled on a nearby roadway.

   _____ h. Motorists traveling through the forest carelessly disposed of lit cigarettes.

   _____ i. Smoke and ash could be seen from miles away.

Lesson 6 Understanding Cause and Effect | 93

Practice Finding Cause and Effect

When traveling through the northwest region of the United States, Elva noticed that the eastern side of a mountain range was very different from the western side. The eastern side was rocky and arid, almost like a desert. The western side was forested and fertile. Elva wondered what had caused these differences.

A. Explain what you think caused these differences. _____

_____

_____

A certain part of the ocean was a favorite place for fishermen to catch lobsters, a shellfish that can be eaten. The fishing was so good that the supply of lobsters in that area was depleted in a few years. Only a few lobsters remained on the ocean floor. Then, a law was passed. This law forbids the taking of lobsters from this area.

B. List some possible effects of this law. _____

_____

_____

_____

In nature, a cause leads to an effect, which results in a series of related causes and effects. This is called a chain of cause-and-effect relationships. Number the following in order from 1 to 6 to show a chain of events.

C. Renumber Items 1 to 6 to show a chain of events. Item 1 should be the cause of Item 2, Item 2 the cause of Item 3, and so on.

_____ 1. Volcanic dust spreads throughout the atmosphere.

_____ 2. Pressure builds up underground.

_____ 3. The polar ice cap increases in size and mass.

_____ 4. A volcano erupts with a tremendous explosion.

_____ 5. Average daily temperature drops two degrees throughout the world.

_____ 6. The dust blocks some of the Sun's warming rays.

▲ **Figure 3-23** A volcanic eruption is both a cause and an effect.

Thinking About Cause and Effect

While investigating cause and effect relationships, you might find new information that could change your ideas about possible causes. New information can also help you select the most probable cause from a list of causes.

Use Figure 3-24 to answer Items 1 to 4. Answer each question completely before going to the next one.

1. Raul noticed that his bicycle tire was flat. Write some possible causes for his tire going flat.

▲ **Figure 3-24** There are many possible causes for this flat tire.

2. Raul looked for a nail or any other sharp object that might be sticking in the tire. He found nothing. Raul pumped air into the tire. The tire went flat again. Now, what are the most likely causes of the tire going flat?

3. Next, Raul put some soapy water on the tire. He saw some air bubbles forming around the bottom of the valve. What do you think is the most likely cause of the tire going flat?

4. How did new information help you determine the most likely cause of the tire going flat?

Extending Your Experience

Answer the following in complete sentences. Use a separate sheet of paper.

1. Find out about the effects of air and water pollution. Are there some effects that do not show up right away? How do scientists establish cause-and-effect relationships between pollution and human illness?

2. Describe at least four different causes of food spoilage. How can you prevent food spoilage? Create a poster to illustrate what you learn.

3. Suggest reasons why the signals from a radio station get weaker as you get farther from the station.

4. What causes the Sun to shine? Everything on Earth would change if the Sun did not shine. Find out the source of the Sun's power. Will it ever stop shining?

LESSON 7 Concluding

Reading for a Purpose
• How do you draw conclusions from observing and studying information?

Terms to Know
conclusion a decision reached about a question under consideration; a final answer or explanation

inference a reasonable conclusion based on information not directly observed

Drawing Conclusions in Science

Once observations are made and information is gathered, scientists draw conclusions. A **conclusion** is a decision reached about a question under consideration. It is a final answer or explanation. One kind of conclusion made by scientists is called an inference. An **inference** is a reasonable conclusion based on information not directly observed.

A. Study Figure 3-25. Then, answer Items 1 and 2 in the spaces provided.

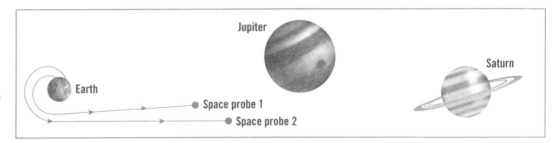

Figure 3-25 ▶ You can make inferences based on information not directly observed.

1. What can you observe about the directions in which the space probes are moving?
   _____

2. What are some possible inferences you can make, based upon your observations?
   _____

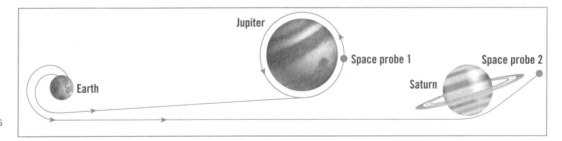

Figure 3-26 ▶ A slight change in a picture can change your conclusions about it.

3. Now look at Figure 3-26. What conclusions can you form based on your observations?
   _____

96 | Chapter 3 Thinking As a Scientist

When scientists are trying to find answers to questions, they collect data and organize it into tables and graphs. Then, they use the tables and graphs to draw conclusions. Drawing conclusions is one of the last steps in trying to answer questions.

B. Read the following scenario. Then, answer Items 4 to 5 using Figures 3-27 and 3-28.

Students experimented to find out how simple machines make work easier. The work was to move a large rock. The simple machine was a 3-m iron bar used as a lever. The units used to measure effort were the number of students needed to push down on the effort arm of the lever in order to overcome the resistance force of the rock. The students changed the position of the fulcrum during the experiment. The results are shown in Figure 3-28.

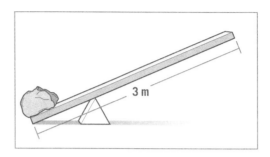
◀ Figure 3-27
Experiment setup

4. How many students did it take to move the rock when the effort arm was

   1.5 m long? _____

   2.5 m long? _____

5. What conclusion, or inference, can you draw about the length of the effort arm and the amount of effort force needed to overcome a given resistance?

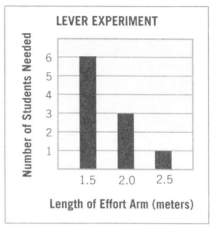
◀ Figure 3-28
Experiment data

_____

The students decided to perform a second experiment. They were going to move the rock using pulleys. Again, the units they used to measure the effort were the number of students needed to overcome the resistance force of the rock. They set up the pulleys as shown in Figure 3-29. The results are given in the chart in Figure 3-30.

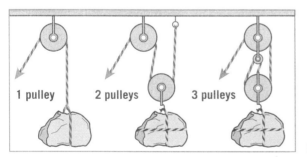
◀ Figure 3-29
Experiment setup

Figure 3-30 ▶
Experiment data

| Pulley Experiment ||
Number of pulleys	Number of students
1	8
2	5
3	3

C. Use the information in Figures 3-29 and 3-30 to answer Items 6 and 7.

6. What conclusion, or inference, can you draw about the use of pulleys and the amount of effort force needed to overcome a given resistance?

_____

7. Compare your conclusions from the two tests. Explain the similarities and differences.

_____

Practice Drawing Conclusions

Suppose you wanted to move to Canada for a while, but you do not know much about the climate. You know that Canada is a large, northern country. Would you expect the climate to be the same all over Canada, or would there be differences? What inferences could you make based on information contained in the chart?

Canada's Climate by Area						
City	Average Temperature (Celsius)		Average Precipitation (milliliters)		Yearly Snowfall (centimeters)	Latitude
	Jan.	July	Jan.	July		
Calgary, Alberta	−11	17	17	68	140	50° 30'N
Halifax, Nova Scotia	−6	18	137	80	160	44° 40'N
Montreal, Quebec	−10	21	76	85	135	45° 30'N
Toronto, Ontario	−4	22	63	81	100	43° 40'N
Vancouver, British Columbia	2	17	147	30	41	49° 30'N

▲ **Figure 3-31** By studying a chart, you can make many inferences.

Use Figure 3-31 to answer Items 1 to 7.

1. Which city has the coldest January? _____

2. Which city has the warmest January? _____

3. Which city has the greatest difference between the average temperatures in January and July? _____

4. Which city has the smallest difference between the average temperatures in January and July? _____

5. Which city has the most snow? _____

6. What conclusion can you make about the climate of Vancouver? Place a check (✓) by the conclusions you draw from the chart.

    _____ a. Vancouver is an oasis in a desert.
    _____ b. Vancouver has warm summers and mild winters.
    _____ c. Vancouver has a marine climate.
    _____ d. Vancouver is cold because it is so far north.

7. Calgary and Vancouver have the same average temperature in July. That average temperature is much less than the average July temperatures of Montreal and Toronto. Considering all the other factors in the table, the following are conclusions you might draw about the locations of the two cities. Place a V by the conclusion you would draw about Vancouver. Place a C by the conclusion you would draw about Calgary.

    _____ a. The city is near an ocean.
    _____ b. The city is near a river.
    _____ c. The city is near a swamp.
    _____ d. The city is inland.

Thinking About Drawing Conclusions

Often there are several factors that should be considered when drawing conclusions. Two of these factors are the length of time over which observations are made and the historical period in which they were made. Ask yourself: Have observations been made or data collected over a long enough period of time? Could new observations or data cause an earlier conclusion to be untrue?

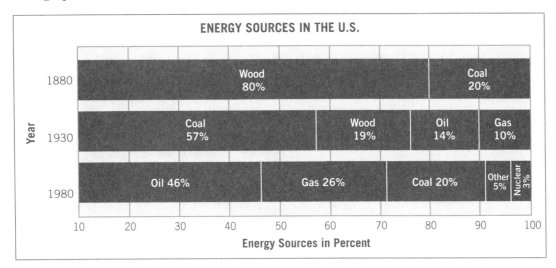

◄ Figure 3-32 Major energy sources over 100 years

Read the following example and answer Items 1 to 3.

Dependency on different types of energy sources in the United States has changed over the years. The figures in the graph in Figure 3-32 represent major energy sources used in the United States over a hundred-year period.

1. What conclusion can you make about the use of oil covered by this data?

2. Why is it important to compare the relative amounts of different energy sources used for different time periods?

3. List another example in which conclusions would depend on the time period being considered.

Extending Your Experience

Answer the following in complete sentences. Use a separate sheet of paper.

1. Look in newspapers or on the Internet for graphs that show trends or give results of studies or experiments. Collect the graphs. Then, draw your own conclusions from the graphs and compare them to the conclusions given in the article accompanying the graph. Try to explain any differences between your conclusions and those in the article.

2. How did scientists draw the conclusion that the continents of the world were once connected? What information did they use to reach this conclusion?

3. Investigate the research of Marie and Pierre Curie or Henri Becquerel. What conclusions were they able to draw from their research? Were any of their conclusions proven wrong by later observations and data?

LESSON 8 Generalizing

Reading for a Purpose
- How do you use observations and patterns of information to make generalizations?
- How can you make inferences using generalizations?

Terms to Know
generalization a broad, general statement about a topic that is true most of the time
valid generalization a generalization that can be supported as reasonable

Making Generalizations

A broad, general statement about a topic that is true most of the time is called a **generalization**. Generalizations can be made when you have a great deal of information about a topic. A generalization is a type of conclusion about everything in a group. A generalization can be made even if you have incomplete data about some of the objects in a group.

A. Study the information in the graph in Figure 3-33. Then, answer Items 1 to 5.

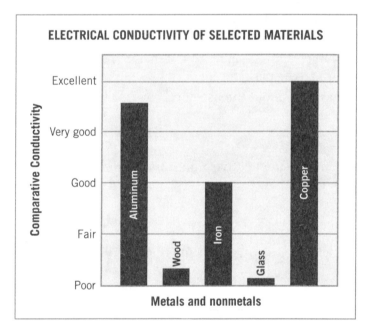

◀ Figure 3-33 A graph comparing electrical conductivity of some metals and nonmetals

1. What does the graph show? _____

2. Based on the data in the graph, the metals listed are

   _____ conductors.

3. Assume that most metals and nonmetals are like the ones listed here. In

   general, then, we can assume that metals are _____

   conductors of electricity. Nonmetals are _____

   conductors.

100 | Chapter 3 Thinking As a Scientist

4. We cannot make generalizations about all substances based on the information in this graph. Why not? _____

5. Generalizations can be used to make plans or to form hypotheses. The ability of metals to conduct electricity compared to the ability of nonmetals to do the same can be very useful information. Put a check (✓) next to each statement that makes use of this information.

   _____ a. what materials to use for the filament of a light bulb

   _____ b. what materials to use in electrical wiring

   _____ c. what materials to use for electrical insulators

   _____ d. what materials to use to make a table

Scientists study many individual objects or situations before making a generalization about them. The generalization then helps the scientists identify other objects of the same kind. An example is the identification of crystal shapes. Many crystals were studied before scientists generalized about their shapes. They classified crystal shapes into six systems. The crystal of each mineral fits one of the six crystal systems.

B. Study the crystal shapes in Figures 3-34 to 3-37. Then complete Items 6 and 7.

6. Figure 3-34 shows a generalized drawing of the cubic system. Circle the minerals in Figure 3-35 that have cubic crystals.

▲ Figure 3-34 This is a cubic shape.

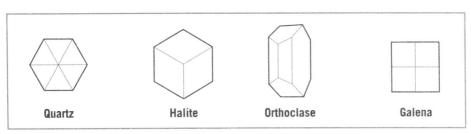

▲ Figure 3-35 Some of these minerals have a cubic shape.

Quartz   Halite   Orthoclase   Galena

7. Figure 3-36 shows a generalized drawing of the hexagonal crystal system. Circle the minerals in Figure 3-37 that have hexagonal crystals.

▲ Figure 3-36 This is a hexagonal shape.

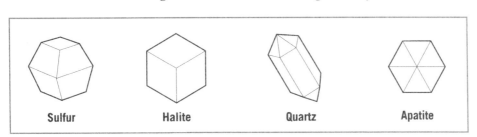

▲ Figure 3-37 Some of these minerals have a hexagonal shape.

Sulfur   Halite   Quartz   Apatite

Generalizations can be used to draw conclusions or inferences about the members of a group. If you know some generalizations about any group, you can infer that the generalizations are true for all members of that group.

C. Read the following generalizations. Then, answer Item 8.

Here are some generalizations about insects.
- Insects have three body parts.
- Insects have three pairs of legs.
- Insects have more species than any animal group.

8. A grasshopper is an insect. Based on the generalizations about insects, what inferences can you make about grasshoppers?

_____

Practice Making Generalizations

Here are some generalizations about the properties of some elements in the periodic table.
- Except for hydrogen, all of the elements located on the left side of the periodic table are metals.
- The elements in the group at the extreme left side of the table are chemically very active.
- None of the active elements are found free in nature. They are always found combined with other elements in compounds.

A. Based on the generalizations about the properties of some elements, answer Item 1.

1. The element lithium, Li, is located in the upper left corner at the extreme left side of the periodic table. What can you infer about lithium?

_____

Here are some generalizations about igneous rocks.
- Igneous rocks form when molten rock cools and hardens.
- Igneous rocks contain crystals.
- In igneous rocks, crystal size depends on rate of cooling. The slower the cooling, the larger the crystals. Rapid cooling at Earth's surface produces very small crystals.

B. Based on the generalizations about igneous rocks, answer Item 2.

2. Obsidian is an igneous rock with very small crystals. Using the generalizations about igneous rocks, what can you say about obsidian?

_____

Thinking About Generalizations

Scientists like to make generalizations. Generalizations can help make sense of the world. However, some generalizations are not valid. A **valid generalization** is one that can be supported as reasonable. You must always be on your guard against generalizations that are not valid. They could prove a block to understanding.

A. Look at the following generalizations. Place a check (✓) by those that you think are *not valid*. One has been done for you.

    ✓   1. Rocks from Kansas are red.

    _____ 2. Crystals have recognizable shapes.

    _____ 3. Soil is brown.

    _____ 4. Glaciers are generally formed in the Northern Hemisphere.

    _____ 5. Mammals have fur.

    _____ 6. Ocean water is cold.

    _____ 7. At the same temperature, humid air feels warmer than dry air.

    _____ 8. Planets have moons.

    _____ 9. Living things are made up of one or more cells.

    _____ 10. Clouds produce rain.

B. In the space provided, explain why the following statement is or is not a valid generalization. *Rivers are long and wide.*

_____

_____

Extending Your Experience

Answer the following in complete sentences. Use a separate sheet of paper.

1. Write one valid generalization for each of the following: glaciers, ocean currents, meteors, snow, and igneous rocks.

2. What is the climate of your area? Is it temperate, marine, desert, or tropical? Make a list of generalizations about your climate. Then, explain whether you think each generalization is valid. Use your own experience as your guide.

3. Look at some models of boats, trucks, or cars. Write a few generalizations you can make about each of these groups of models.

4. Find exceptions to the rule for common generalizations about any topic. Use topics such as chemistry, appliances, or trains, or choose your own topic.

5. Make a list of generalizations about humans, crustaceans, and bryophytes.

Lesson 8   Generalizing

Chapter 3 Review

Concept Review

A. Compare a lake and an ocean.

1. In what ways are they similar? _____

   _____

2. In what ways are they different? _____

   _____

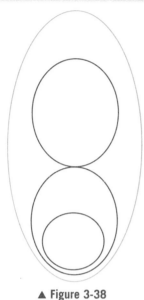

▲ Figure 3-38

B. Classify these groups of objects. Label the diagram in Figure 3-38 to show how the groups are related to each other.

3. gases   4. poisonous gases   5. oxygen   6. nonpoisonous gases

C. Figure 3-39 shows the cyclical pattern of the Moon's changes in shape month after month. Draw in the missing shapes in 7 and 8 to complete the pattern.

Figure 3-39 ▶ This illustration shows the phases of the Moon each month.

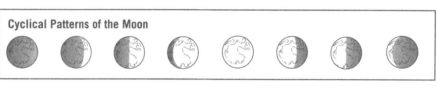

7. _____   8. _____

D. Circle the cause and underline the effect in Items 9 to 12.

9. Movement along a fault line produces an earthquake.

10. Magnetic storms on Earth are caused by solar particles entering Earth's magnetic field.

11. Hail is formed when raindrops pass through layers of cold air.

12. Winds moving across the ocean surface create ocean currents.

E. Read the following scenario. Then, answer Item 13 in the spaces provided.

   When Sarita throws a ball, she tracks how far it goes. She throws it with the same force each time. She finds that when she throws the ball at a very low angle or a very high angle, the ball does not travel very far. When she throws the ball up at about a 45° angle, it goes farther. What inference or conclusion can she draw?

13. _____

   _____

F. Answer Items 14 and 15 in the spaces provided.

14. Write a generalization about the changes that a substance goes through as it moves from the gas phase to the solid phase.

   _____

15. Describe a substance that does not fit your generalization. _____

Vocabulary Review

Write the letter of the Term to Know from the list that best matches each description.

a. compound	d. subgroup	g. inference	j. classify	m. pattern
b. chemical symbol	e. chemical formula	h. effect	k. comparison	n. relationship
c. subscript	f. generalization	i. prediction	l. conclusion	o. cause

_____ 1. a recognizable and reliable design

_____ 2. a notation that shows the kinds and numbers of atoms in a substance

_____ 3. an event or condition that brings about an action or result

_____ 4. to arrange objects into groups based on likenesses

_____ 5. a substance made up of two or more elements that are chemically combined

_____ 6. a decision reached about a question under consideration

_____ 7. when classifying groups, a group that is smaller than a larger one

_____ 8. a natural connection between two objects or events

_____ 9. a shortened way of writing the names of elements

_____ 10. a principle drawn from the study of many individual objects or events

_____ 11. a reasonable conclusion based on the evidence at hand

_____ 12. an examination of similarities and differences

_____ 13. an expressed opinion about what is to come based on some degree of special knowledge

_____ 14. an action or result of a cause

_____ 15. the number used to show how many atoms are present in a chemical formula

CHAPTER 4

Communicating As a Scientist

In the three previous chapters, you learned such skills as how to observe, investigate, predict, sample, classify, and record data. In this chapter, you will learn how to communicate as a scientist.

A scientific discovery is of little value if the scientists who make the discovery cannot communicate it to others. The scientific report is an essential part of scientific progress. Reports make their way around the world. One year, an American scientist may write a report that helps a Japanese scientist in her research. The next year, the opposite may be true.

There are guidelines to be followed in compiling a report. A scientist must use scientific terminology. Illustrations and graphs can be used to make the report clear and complete. Reports written in a specific format are very helpful. Scientists can take these reports and investigate further. Such a system of communication has led to many scientific discoveries.

SKILLS	
✦ Using Science Vocabulary	LESSON 1
✦ Using Illustrations and Models	LESSON 2
✦ Making and Using Graphs	LESSON 3
✦ Writing Scientific Reports of Experiments	LESSON 4
✦ Using Technology to Communicate	LESSON 5

▲ Figure 4-1 A global positioning satellite (GPS)

COMMUNICATING data and information is very important to all scientists. One way of receiving information about Earth and its changes is a global positioning satellite (GPS.)

After the Northridge earthquake in 1994, this scientist from the Jet Propulsion Laboratory examined data at the GPS site at Oat Mountain. Dr. A. Donnellan found that the San Andreas Fault had shifted considerably.

Are you familiar with any other uses of the GPS?

LESSON 1 Using Science Vocabulary

Reading for a Purpose
- How can you identify word parts in scientific terms?
- How can word parts be used to help you understand scientific terms?
- How are some scientific terms created from common words?

Term to Know
terminology (as related to a science) the total of all the special words, or terms, that are used in that branch of science

Understanding Through Terminology

Scientists need to be clear and exact in the words they use. Clearness is especially important when naming things—and there are many things that need exact scientific names. For example, a person looks at the night sky and sees stars. However, a scientist looks at the sky and sees many different kinds of objects, including comets, planets, novas, constellations, as well as stars. Part of becoming a scientist is learning such terminology. **Terminology**, as related to a science, is the total of all the special words, or terms, that are used in that science. Scientists must learn the terminology specific to the field in which they are working.

You will come across many new science words as you study the different branches of science throughout middle school and high school. By knowing the word parts—prefixes, suffixes, and roots—you can usually figure out the meanings of the words. The same word parts are used in many different words, but their meanings are always the same. It is easy, then, to determine the meanings of many science words. For example, the word *hydrology* has two parts: *hydro-*, which means "water," and *-ology*, which means "study of," or "science of."

A. Based on the meanings of the word parts, answer Item 1 in the space provided.

  1. What does *hydrology* mean? _____

B. Study the list of word parts in Figure 4-2. Then, use the list to define the science terms in Items 2 to 14. Check your definitions against those in a dictionary.

Science Word Parts	
Prefixes	**Suffixes**
archaeo- ancient, primitive	*-derm* skin, covering
astro- stars and outer space	*-graph* draw, record
chloro- green	*-lateral* side
chrom- color	*-meter* measuring device
endo- inside, within	*-nomy* dealing with knowledge about
hemi- half	*-oid* like, resemble
micro- very, very small	*-pathy* feeling
pale-, paleo- dealing with ancient conditions or forms	*-plast* cell; cellular particle
poly- many	*-scope* see, examine
therm- heat	*-sphere* field or area
uni- one	*-tropic* turning or changing in response

Figure 4-2 ▶
Table of word parts

2. endothermic _____

3. polylateral _____

4. chloroplast _____

5. chromatograph _____

6. thermometer _____

7. archaeology _____

8. astronomy _____

9. paleontology _____

10. hemisphere _____

11. microscope _____

12. asteroid _____

13. unilateral _____

14. Now, use some word parts to create your own words. Then, define them.

_____

Practice Using Science Vocabulary

Scientific terms are descriptive. Often a scientific term is a combination of words we all know. You know that a continent is a large land mass. Scientists put the word *continental* in front of common words to make scientific terms that apply to parts of continents. You also know these words: *crust*, *rise*, *shelf*, and *slope*. When used with *continental*, these words form scientific terms.

- The *continental crust* is the part of the crust that makes up the continents. It reaches all the way to the far edge of the continental rise.
- The *continental shelf* is the fairly flat part of the continental crust just off the shore of the continent.
- The *continental slope* is the part of the continental crust that slopes steeply down from the shelf to the continental rise.
- The *continental rise* is the part of the continental crust that extends from the base of the continental slope to the deep ocean floor.

A. Use the terminology you have learned to label Figure 4-3. Print the terms on the lines numbered 1 to 3. The continental margin begins at the base of the continental rise and goes to the water's edge. Add this label to the picture and draw a bracket to show how far the continental margin extends.

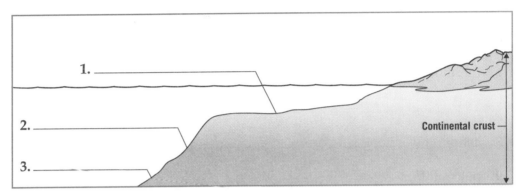

Figure 4-3 ▶ Labels are important to the information in an illustration.

4. How many of the features shown in Figure 4-3 are included in the continental margin? _____

_____

5. What do you think *continental drift* means? _____

_____

B. Here are some more common science word parts and their meanings. Use these and the earlier word part meanings to define the terms in Items 6 to 10. Check your definitions with those in a dictionary.

Science Word Parts	
Prefixes	**Suffixes**
am- ampere	*-lysis* freeing, loosening
electro- electricity	*-metry* measuring
photo- light	*-sonic* relating to sound
ultra- more than, excessive, beyond	*-stat* steady, constant

▲ Figure 4-4 Common science word parts

6. thermostat _____

7. electrolysis _____

8. ammeter _____

9. photometer _____

10. ultrasonic _____

Chapter 4 Communicating As a Scientist

Thinking About Using Science Vocabulary

You can use what you already know about science word parts to help you identify new science words.

A. For Items 1 to 3, define the unknown word using the meanings of the words you already know. Circle the parts of the words you used to define the unknown word.

1. Geology is the study of Earth. A chemist studies matter. What does a geochemist do? _____

2. Biology is the study of life. Physics is the study of the interaction of matter and energy. What is biophysics? _____

3. A thermometer measures heat. Hydrodynamics is the study of the energy of moving water. What is thermodynamics? _____

B. Answer Items 4 to 7 in the spaces provided using the following information. Geophysics is a field of scientific study.

4. Write the parts of the word *geophysics*. _____

5. Give the meaning of each word part. _____

6. What does *geophysics* mean? _____

7. How can you check your definition? _____

Extending Your Experience

Answer the following in complete sentences. Use a separate sheet of paper.

1. Make a list of ten science words. Separate each word into its parts, including prefixes, suffixes, and roots. Use a dictionary to find out how each word came into being. Then, try to make five new words from the parts.

2. Look up the word *terminology* in a dictionary. What parts make up that word? What do the parts mean?

3. Many scientific terms are made up of words from Latin or Greek. For centuries, those languages were the languages of the educated. Today, many scientists take courses in Latin or Greek in order to understand scientific terminology better. Find out if those languages are taught in your school system. Look up those languages in any encyclopedia and study their histories.

LESSON 2

Using Illustrations and Models

Reading for a Purpose
- How can you communicate information and ideas using illustrations or models?
- How do you interpret cross-sectional views of objects?
- How do you use and make flowcharts that show how things are organized?

Terms to Know

illustration a drawing that communicates information, helping to make something clear

label words used in an illustration to describe a part of that drawing

cross section an illustration of what an object would look like if part of it were cut away to show what is inside

longitudinal section a cross section that is cut through the long axis of something

transverse section a cross section at a right angle to a longitudinal section

scale a distance used on a map to represent a greater distance on Earth's surface

contour interval the difference in altitude between two adjacent lines on a topographic map

Interpreting and Drawing Illustrations

An **illustration** is a drawing that communicates information. Often scientists draw illustrations to explain observations and hypotheses. Such a picture or diagram helps make something clear. Information can be shown in a compact, clear, and precise way through illustrations. Scientists use many kinds of illustrations. These include labeled figures and many kinds of maps. An illustration need not include all the details one sees in nature. Usually, only the most important parts of what is being illustrated are shown. **Labels** are used in illustrations to explain the purpose of the picture or diagram.

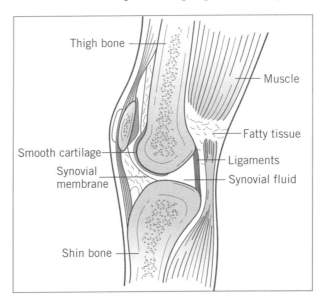

◀ **Figure 4-5** This illustration identifies parts of a human leg.

A. Using Figure 4-5, answer Items 1 and 2 in the spaces provided.

1. What does this illustration show? _____

2. What do the labels communicate as important in this illustration? _____

_____

112 | Chapter 4 Communicating As a Scientist

Here are some rules to follow when drawing illustrations.
- Keep the drawing as simple as possible.
- Keep the objects you draw in proportion to one another.
- Print the labels clearly. Write them horizontally.
- Draw straight lines from the labels to the items you are labeling.

B. Study a common object, such as a pencil or pen. Decide what its main features are. Draw an illustration of it in this space. Label the features and write a caption.

To show what the inside of something looks like, a special kind of illustration is sometimes used. A **cross section** is an illustration of what an object would look like if part of it were cut away to show what is inside. There are two types of cross sections. A **longitudinal section** shows a cross section that is cut through the long axis of something. A **transverse section** shows a cross section at a right angle to a longitudinal section.

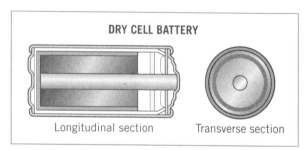

▲ **Figure 4-6** This drawing shows two types of cross sections of a dry cell.

C. Label each cross section in Items 3 to 6 as a longitudinal section or a transverse section.

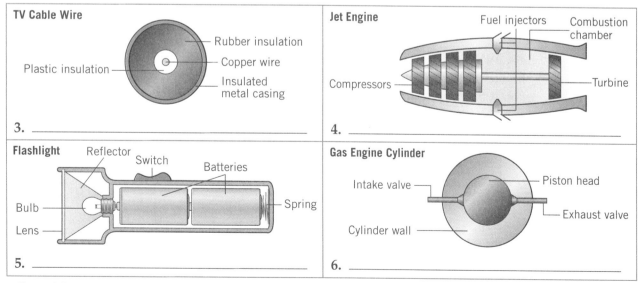

3. _____

4. _____

5. _____

6. _____

▲ **Figure 4-7** Each of these items shows a different cross section.

Lesson 2 Using Illustrations and Models

Scientists use many kinds of maps. An important rule to remember about all maps is that they are drawn to scale. A **scale** in this sense is a distance used on the map to represent a greater distance on Earth's surface.

A topographic map is a special type of map that shows altitude by using a contour interval. A **contour interval** is the difference in altitude between two adjacent lines on a topographic map.

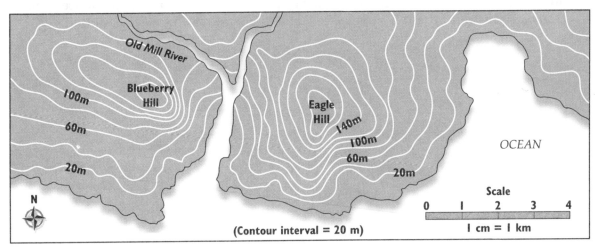

▲ **Figure 4-8** This is a contour map.

D. Use the contour map in Figure 4-8 to answer Items 7 to 9.

7. What is the scale of this map? _____

8. Using a metric ruler, measure the distance in centimeters between the top of Eagle Hill and the top of Blueberry Hill. What is the distance, in kilometers, between these two points? _____ km

9. The contour interval of this map is 20 m. Starting at the bottom of each hill, add 20 m for each contour line you cross on your way to the top. Which is higher, Eagle Hill or Blueberry Hill? How high is each hill?

_____

Practice Making Illustrations

Flowcharts show steps in a process. A flowchart is another way to give information about the ways parts are organized within a large system.

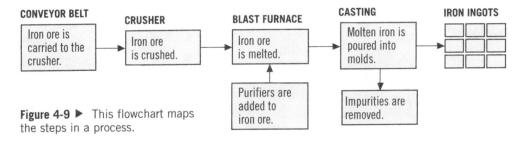

Figure 4-9 ▶ This flowchart maps the steps in a process.

114 | **Chapter 4** Communicating As a Scientist

Study the flowchart in Figure 4-9. Then, on a separate sheet of paper make a flowchart to illustrate the proper sequence of events by which electricity is generated in a hydroelectric power plant. The events are listed below.

    Water is directed onto turbines.

    Electricity leaves the power plant and is sent to customers.

    Generators change mechanical energy into electricity.

    Water falls from a higher level to a lower one.

    Turbines turn electrical generators.

Thinking About Using Illustrations

In scientific reports, many ideas are expressed in the form of illustrations. Illustrations are often better for communicating some types of information than the written word.

Items 1 to 7 are some typical reasons for using illustrations in science reports. Put an *R* next to the reasons that make illustrations helpful to the reader of a report. Put an *A* next to those reasons that an author might use for putting illustrations in reports. Some reasons may be useful to both reader and author.

  _____ 1. It takes less time to look at an illustration than to read a description.

  _____ 2. Illustrations show what things look like.

  _____ 3. Illustrations show what the author thinks is important.

  _____ 4. Illustrations are easily understood.

  _____ 5. Illustrations can express new ideas and show trends in data.

  _____ 6. Illustrations save space.

  _____ 7. Illustrations make reports more interesting.

        8. What might be another reason for using illustrations in science reports?

    _____

    _____

Extending Your Experience

Answer the following in complete sentences. Use a separate sheet of paper.

1. Find out what a cartographer is, what he does, and what training a cartographer needs for his job.

2. Compare hand-drawn illustrations to those drawn by computers. List the similarities and differences. List some advantages and disadvantages of each.

3. Often a photograph looks better at first glance than a labeled illustration. Aerial photographs (taken from airplanes) are especially useful. Why do we not always use photographs to explain and compare? In what ways can illustrations be more useful than photographs?

LESSON 3 — Making and Using Graphs

Reading for a Purpose
• How do you read bar graphs, line graphs, and circle graphs?
• How can you make graphs from collected information?

Terms to Know

graph a visual way to communicate information

bar graph a graph that uses bars of different lengths to show relationships; the bars can be horizontal or vertical

axis a straight line used for reference

vertical axis the upright axis of a graph

horizontal axis the crosswise axis of a graph

line graph a graph that uses lines to show trends or changes over time

circle graph a circular diagram that shows relative sizes of the parts that make up the whole

interpolation a prediction made for a value between observed data

extrapolation a prediction made for a value beyond observed data

Using Graphs

Scientists collect information. Then, they find ways to organize the information. Graphs are helpful tools. A **graph** is a visual way to communicate information. One type of graph is a **bar graph**. This type of graph uses bars of different lengths to show data. The bars can be placed horizontally or vertically.

Graphs are usually constructed with two main lines called axes. An **axis** is a straight line used for reference. The two axes of a graph often meet at zero. The **vertical axis** is the upright axis. The **horizontal axis** is the crosswise axis.

A. Study the bar graph in Figure 4-10. Then, answer Items 1 to 6.

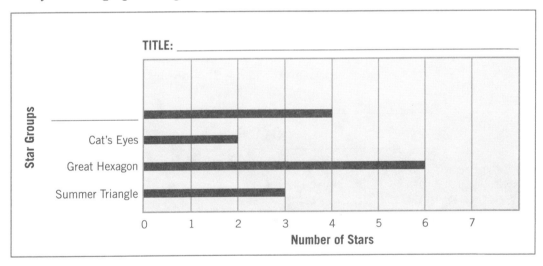

Figure 4-10 ▶ This bar graph shows star groups.

1. Write a short, descriptive title in the space above the graph.

2. What do the horizontal axis and the vertical axis on the graph show?

_____

3. This bar graph lacks a label, the star group that has a bar indicating four stars. It is called the Great Square. Add that label to the bar graph in Figure 4-10.

116 | Chapter 4 Communicating As a Scientist

4. Which star group has the fewest number of stars? _____

5. Which star group has the greatest number of stars? _____

6. A star group called the Big Dipper has seven stars. Add a label and a bar to the graph to show this information.

Information is often organized in a chart or table. A graph of that information may be made to show the data in a different way.

B. Using the set of axes to the right, draw vertical bars to display this data in a graph. Put numbers on the vertical axis and element names on the horizontal lines. Write a title for the graph.

Number of Protons in Atoms	
Element	Protons
Helium He)	2
Boron (B)	5
Nitrogen (N)	7
Oxygen (O)	8

▲ **Figure 4-11** You can use the information in a table to create a graph.

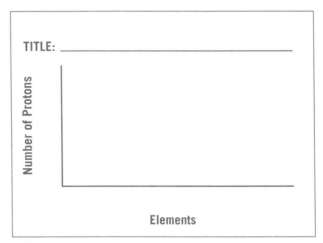

▲ Figure 4-12

C. A graph can be very helpful when comparing two or more things. Figures 4-13 and 4-14 show types and numbers of fossils found in two different locations. Answer Items 7 and 8. Then, complete the graph in Figure 4-14 by adding gastropods.

Types of Fossils and Locations		
Type of Fossil	Found in Canyon	Found in Cave
Trilobites	15	22
Brachiopods	30	12
Gastropods	26	5

▲ Figure 4-13 Fossil locations

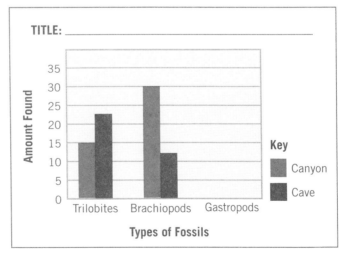

Figure 4-14 ▶

7. What things are compared in the table? _____

8. Explain why two kinds of shading are used on the bars.

_____

Lesson 3 Making and Using Graphs | 117

Another type of graph is a line graph. **Line graphs** use a line or lines to connect data points. The lines show trends or changes over time. The line graph below was made using the information in the table.

A line graph can be used for estimating information. For example, you could estimate the amount of snow that was on the ground on January 15. To begin, find January 15 on the horizontal axis. Then, follow a straight line from that point up to the graphed line. Place an X at that spot. Trace a horizontal line from the X, moving left to the vertical axis. How many meters does the vertical line read at the place your line intersects with the vertical axis? Thus, you can estimate that on January 15, there were 4 m of snow on the ground in Buffalo. For other dates that do not appear on the horizontal axis, just estimate about where they would be located. Then, follow the same process to find the figure for the amount of snow on the vertical axis.

D. Use the information in Figures 4-15 and 4-16 to answer Items 9 and 10.

Amount of Snow on the Ground in Buffalo, NY	
Date	Snow (meters)
Dec. 15	1
Jan. 1	2
Feb. 1	6
Feb. 15	8

▲ **Figure 4-15** Snowfall data

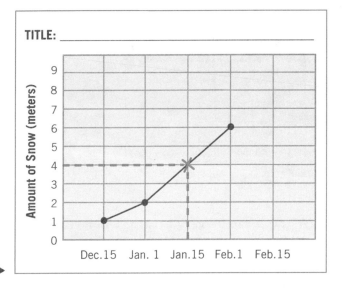

Figure 4-16 ▶

9. How much snow was on the ground on February 15? Finish the line graph by graphing the figure for that date. Be sure to draw a line between the dot for February 1 and the dot you make for February 15.

10. How many meters of snow were on the ground on these dates? Add a dot for each of the dates. The first one is done for you.

   a. January 15 __4 m__   b. January 8 _____   c. February 8 _____

Another type of graph is a circle graph.
A **circle graph** shows the relative sizes of the parts that make up the whole. It is sometimes called a pie graph because it shows the sizes of the pieces that make up *the whole pie*.

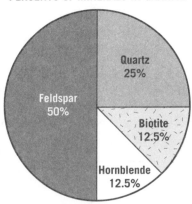

Figure 4-17 ▶ This circle graph shows all the minerals in granite.

118 | Chapter 4   Communicating As a Scientist

E. Use the information in Figure 4-18 to create a circle graph in Figure 4-19.

Gases in the Atmosphere	
Gas	Percent
Oxygen	21.00
Carbon dioxide	0.04
Nitrogen	78.00
Water vapor, helium, and other gases	0.02
Argon	0.94

▲ Figure 4-18 Percents of gases in the atmosphere

TITLE: _____

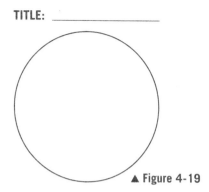

▲ Figure 4-19

F. There are many types of graphs. When you want to graph some information, you have to decide which of the graph types would best display your information. Study the two graphs in Figures 4-20 and 4-21. Then, answer Items 11 to 15.

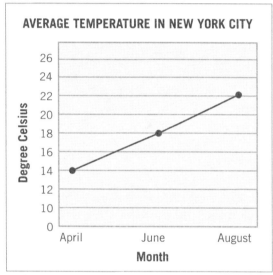

▲ Figure 4-20 Temperature data for New York City

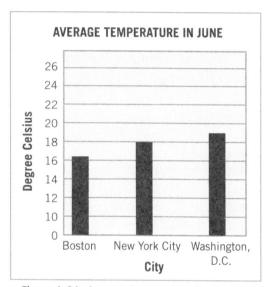

▲ Figure 4-21 Average temperature for three cities

11. Which graph compares temperatures in different cities? _____

12. Estimate of the average temperature in New York City during May. _____

13. Would either graph be helpful in estimating the average temperature in New York City during January? Explain your answer. _____

14. During the month of June, what is the difference between the average temperatures in Boston and in Washington, D.C.? _____

15. Would either graph be helpful in estimating the average temperature of Phoenix, Arizona, during June? Explain your answer. _____

Lesson 3 Making and Using Graphs | 119

Practice Using Graphs

Remember these steps when you are drawing your own graph:

Step 1 Display the data in a table.
Step 2 Draw and label the vertical and horizontal axes.
Step 3 Space the numbers and labels evenly to fit the entire length of each axis.
Step 4 Arrange the graph neatly so it can be understood easily.
Step 5 Give the graph a title.

Brianna was writing a report on elements found in living things. She was required to include at least two visuals. Brianna chose to create a chart of information and then graph it.

Elements Found in Living Things	
Elements	Percents
Oxygen (O)	64.5
Carbon (C)	18.0
Hydrogen (H)	10.0
Sulfur (S), phosphorus (P), and others	4.5
Nitrogen (N)	3.0

A. Figure 4-22 contains information Brianna found. Decide the type of graph that best communicates the information. Then, draw it on a separate sheet of paper. Be sure to follow the steps above.

▲ **Figure 4-22** A chart of elements found in living things

Jiro and Alida investigated the rates at which a flat sheet of paper and a wadded sheet of the same kind of paper fell to the ground. Trials were run from several different heights. In each trial, the two items were dropped from the same height. The time it took each item to reach the ground was measured.

B. The data for the wadded paper has been graphed in Figure 4-24. Add the data for the flat sheet of paper to the graph. Connect the points with a dashed line. Draw a key for the two lines. Give the graph a title.

Comparison of Rates of Fall		
Height (meters)	Time of Fall (seconds)	
	Flat sheet of paper	Wadded paper
2	3	.5
4	6	1.0
6	10	1.5
8	15	2.0
10	20	2.5

▲ **Figure 4-23** Data for falling paper

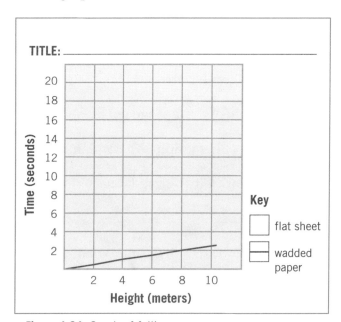

▲ **Figure 4-24** Graph of falling paper

Thinking About Using Graphs

Graphs can be helpful in making fairly accurate estimations and predictions. An **interpolation** is a prediction of a value that falls between observed data.

An **extrapolation** is a prediction of a value made beyond observed data. An extrapolation is made by imagining what the graph would look like if it were continued. When making an extrapolation from a line graph, look for a pattern in the points graphed. Is the line going up or down at the end? Does the line go up or down in a regular way? Those are two questions you might ask yourself.

Look at the graph in Figure 4-25 of a hurricane's position relative to land over a three-day period. Scientists graph a hurricane's positions to help them predict where and when the hurricane will hit land.

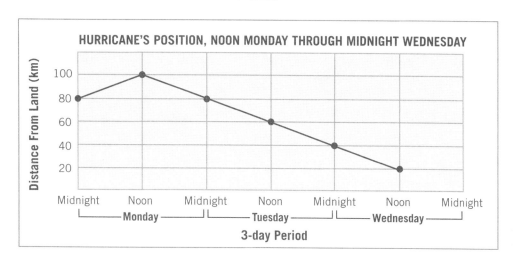

◀ Figure 4-25
Collecting data about a hurricane helps scientists predict when the hurricane will hit land.

1. Make an interpolation of the hurricane's distance from land on Tuesday at 6:00 P.M. Look back to page 118 for help. _____

2. Make an extrapolation to predict when the hurricane will hit land. Hint: Look for a pattern in the graphed line that will help you make a prediction.

Extending Your Experience

Answer the following in complete sentences. Use a separate sheet of paper.

1. Symbols are often used to represent words or numbers. Pictographs are symbols used to display data. Find examples of bar graphs in magazines and books. Make up symbols for the labels in the graphs and make pictographs.

2. Make a circle graph of the colors you are wearing today. First, write down all the colors you find on your clothes and shoes. Then, estimate what percent of the total is made up by each of the colors on your list. Finally, use these percents to make your pie. Be sure that each of the pieces accurately represents how much of that color you are wearing. Label the different sections. Use colored pencils or pens to shade the sections.

Lesson 3 Making and Using Graphs

LESSON 4 Writing Scientific Reports

Reading for a Purpose
- What information should be included in a formal scientific report?
- How should a formal scientific report be organized?
- How do you write a report from a description of an experiment?

Term to Know
report a detailed account of something

Reporting Experiments

Part of what scientists do is to report their findings to other scientists. **Reports** are detailed accounts of something. Written reports usually have several sections. The following is an example of a student's report of a chemistry experiment.

TITLE: Comparing Chemical Reactivity of Two Acids—Hydrochloric (HCl) and Acetic (CH_3COOH)

HYPOTHESIS: Acetic acid is weaker (less reactive) than hydrochloric acid.

INTRODUCTION: Acids are corrosive. They tend to break down certain substances. Some acids are stronger than other acids. Stronger acids are more reactive. The purpose of this experiment is to compare the relative strengths of the two acids.

MATERIALS: Chalk, dilute (5%) solutions of (CH_3COOH) and HCl, 2 plastic beakers, 2 forceps, graduated cylinder, balance, safety goggles, gloves

Safety Note: Handle acids with extreme care. They can cause painful burns. Wear goggles and gloves.

PROCEDURE:
1. Measure out 100 mL of 5% HCl solution and pour it into one beaker.
2. Rinse the graduated cylinder. Measure out 100 mL of 5% CH_3COOH solution and pour it into the second beaker.
3. At the same instant, carefully place a 5 g piece of chalk (calcite) into each beaker. Avoid splattering the acid.
4. After 15 minutes, use forceps to remove the pieces of chalk from the beakers. Both pieces of chalk should be removed at the same time.
5. Thoroughly rinse and dry both pieces of chalk. Then, measure the mass of each piece of chalk on the balance.

RESULTS/OBSERVATIONS:
1. The mass of the chalk that was in the HCl solution is 3 g. The chalk "lost" 2 g of mass reacting with HCl.
2. The mass of the chalk that was in the CH_3COOH solution is 2.5 g. This piece of chalk "lost" .5 g of mass reacting with CH_3COOH.

CONCLUSION: CH_3COOH is a weaker (less reactive) acid than HCl.

▲ Figure 4-26 This is a sample of report writing.

A. Read the report in Figure 4-26 and answer Items 1 to 8.

1. List the seven sections of this report. _____

   _____

2. Which section explains the purpose of the experiment? _____

   _____

3. Which section states what is to be tested? _____

4. Which section describes the steps followed in performing the experiment?

   _____

5. Which section lists the supplies and equipment used in the experiment?

   _____

6. Which section gives the data that was collected? _____

7. Which section compares the results with the hypothesis? _____

8. Why would a person preparing such a report include a safety precaution

   about acids? _____

   _____

Reports should contain specific, clearly written information. Nothing should be left to the imagination of the reader.

B. Consider each of the following pairs of statements. Place a check (✓) by the statement in each pair that you think would be proper for a report.

9. **Title:**   _____ Batteries and Lights

                        _____ Effect of Increasing Voltage on Lamp Brightness

10. **Hypothesis:**   _____ The brightness of the light bulb is directly related to the number of batteries in a circuit.

                        _____ More batteries will give a brighter light.

11. **Procedures:**   _____ Add a second battery to the circuit.

                        _____ Hook up more batteries.

12. **Results/
    Observations:**   _____ The light bulb was brighter with more batteries.

                        _____ When a second battery was added to the circuit, the light bulb was twice as bright as it was with one battery.

Lesson 4 Writing Scientific Reports | 123

C. Graphs, charts, and illustrations can help make a report clearer to the reader. Write the section in which each of the following would appear.

13. a chart of information collected during the experiment _____

14. an illustration of the way a material is to be held _____

15. a drawing of the apparatus to be used _____

16. What kind of illustration might have been helpful in the report on page 122?
_____

Practice Reporting Experiments

Read the following description of an experiment. Then, write a report of the experiment. Include the seven sections you listed on page 122. You may start your report in the spaces provided. If needed, use another sheet of paper.

Wing and Chi know that a barometer is used to measure air pressure and that barometer readings change with the weather. They decided to conduct an investigation. Their hypothesis was that a rapidly falling barometer indicates the approach of stormy weather. Their apparatus included a clock, an aneroid barometer, a pencil, and paper.

First, they made a chart to record information. Second, they checked the barometer six times a day—every hour on the hour beginning at 10:00 A.M. and ending at 3:00 P.M. Each time they recorded the barometer readings. They also observed and recorded the weather conditions each hour. They continued their investigation for 3 weeks.

The results were expected. Some of the data they collected follows.
- On days 1 through 6 of the first week, the barometer readings held steady. The weather was fair.
- On day 7, however, at the end of the first week, the barometer fell rapidly. It was 1:00 P.M. By 4:00 P.M., there was a thunderstorm.
- Wing and Chi noticed similar activity on day 13 of the second week and day 17 of the third week. Rainy weather appeared by 12:30 P.M. on the thirteenth day. There was a severe thunderstorm at 5:00 P.M. on the seventeenth day.
- The weather on the remaining days was fair and pleasant.

_____
_____
_____
_____
_____
_____
_____
_____

Thinking About Reporting Experiments

Scientists must communicate with each other in ways that other scientists can understand. The language that scientists use must be clear and concise. Sometimes sentences can be misleading or have more than one meaning. This can confuse the reader.

Read each of the sentences in Items 1 to 3. Then, tell what is confusing about each. Rewrite each sentence to make the meaning clear.

1. We watched Juan use a lever to lift a rock. It broke.

   a. What is confusing? _____

   b. Rewrite the sentence to make the meaning clear. _____

2. Maria compared the boiling point temperature of water to the boiling point temperature of alcohol. It was higher.

   a. What is confusing? _____

   b. Rewrite the sentence to make the meaning clear. _____

3. The samples of carbon dioxide and oxygen were cooled until one of them became a solid.

   a. What is confusing? _____

   b. Rewrite the sentence to make the meaning clear. _____

Extending Your Experience

Answer the following in complete sentences. Use a separate sheet of paper.

1. Scientists must be especially careful to record their procedures so that other scientists can repeat their experiments. Take a simple, everyday task and list the procedures you follow to complete that task. Make sure your description is clear and complete. Check your description by having a friend follow your procedures to do the task.

2. Many science magazines publish reports of recent investigations. Such articles are usually not written in the form of a formal report. Find such an article. Check to see if the article contains all the information you would need to write a formal scientific report.

3. Suppose you are a scientist who has conducted many successful experiments. In one experiment, the results were inconclusive. *Inconclusive* means that you were not able to draw any definite conclusions. Would a report on this experiment be worth writing? Could you write a report and say your findings were inconclusive? Explain.

4. Some scientists consider the Procedures section to be the most important part of a scientific report. Do you agree? Explain your answer.

LESSON 5 — Using Technology to Communicate

Reading for a Purpose
• How do scientists use technology to communicate?
• What are some ways you can communicate about science using technology?

Terms to Know
technology the application of scientific knowledge to produce things needed or desired by people
Internet a network that allows computers around the world to communicate with each other
wired technology technology that sends information through wires
wireless technology technology that sends information from place to place without the use of wires

Types of Communication Technology

Scientists need to share ideas about science with each other. Advances in technology have helped scientists, as well as others, to communicate. **Technology** is the application of scientific knowledge to make it useful to people.

Personal computers are one way technology has helped scientists to communicate. Personal computers use software to do many things. Spreadsheet software allows scientists to make graphs and charts. Word-processing software allows them to produce written reports. Presentation software allows them to use their personal computer during a presentation. Scientists can also use the personal computer to connect to the Internet and do research or send messages electronically. The **Internet** is a network that allows computers around the world to communicate with each other. In fact, the Internet was created to make it easier for scientists everywhere to communicate.

A. Match one of the following uses of the personal computer with the scientist's need. Choices may be used more than once.

a. presentation software

b. spreadsheet software

c. word processing software

d. Internet

e. electronic mail

_____ 1. create a chart comparing global ocean temperatures

_____ 2. write a report on the feeding habits of toucans in Belize

_____ 3. graph the growth of corn seedlings

_____ 4. present information on new cancer treatments

_____ 5. research the mineral composition of local soil

_____ 6. invite a colleague to a lecture on white dwarf stars

_____ 7. type the procedure for a new experiment

Chapter 4 Communicating As a Scientist

Personal computers that connect to the Internet through wires are an example of a wired technology. **Wired technology** sends information through wires. **Wireless technology** sends information from place to place without the use of wires.

Scientists also use wireless technology to communicate. Weather satellites communicate information about Earth's atmosphere to computers that help meteorologists predict weather. Other satellites allow doctors in one hospital to study the X-rays of a patient in another hospital far away. Wireless technology is also being used to help scientists search for other forms of intelligent life in the universe.

B. Tell whether each of the following types of technology is wired or wireless by writing *wired* or *wireless* in the space provided.

_____ 8. weather satellite   _____ 10. cell phone   _____ 12. cable television

_____ 9. personal computer   _____ 11. radio

Scientists are not the only ones who use technology to communicate about science. Just like scientists, students use personal computers to create graphs and charts, write reports, create presentations, send messages electronically, and use the Internet. Students also use computers and calculators to collect data during experiments and to communicate with each other. Businesses and emergency services, too, find this technology an important part of their daily operations.

C. Figure 4-27 shows different technological tools used for communication. Refer to it to respond to Items 13 to 17. Some items may have more than one answer.

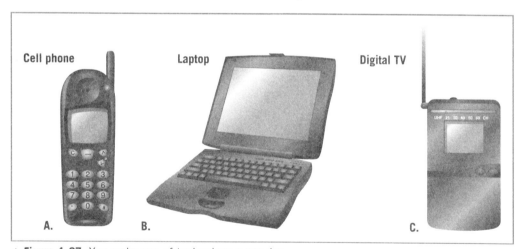

▲ **Figure 4-27** You make use of technology every day.

13. Which tool would you use to call someone and say that you need a ride home from school? _____

14. Which tool could be used to connect to the Internet? _____

15. Which tool lets you watch television while outside? _____

16. Which tool allows you to send electronic mail to a friend? _____

17. Which tool allows you to type a paper? _____

Practice Using Technology

A scientist conducted a study of the changes in a population of owls in a forest over a period of 10 years. The chart in Figure 4-28 contains the results of her study. On a separate sheet of paper, prepare a graph of her data. If available, use a spreadsheet program or a graphing calculator to plot the data.

Owl Population Over a 10-Year Period										
Year	1	2	3	4	5	6	7	8	9	10
Owl Population	25	33	31	28	32	36	46	52	61	75

▲ **Figure 4-28** Changes in owl population over a 10-year period

1. Over the 10-year year period shown, did the population get larger or smaller in size? Was this change in size constant over the entire 10 years? Explain.

2. During what years did the owl population decrease in size? Suggest a reason that the population of owls might have decreased.

3. During what year did the population grow the most? Suggest a reason why the population of owls might have increased so much in one year.

4. In what ways could this scientist use the information from her experiment to communicate the health of this owl population to other scientists?

5. What factors might cause the owl population to change in size from year to year?

Thinking About Technology

Read each description of a way used to communicate information. In the spaces provided, describe how you could use today's technology to communicate the same information.

1. sending a letter by Pony Express

2. typewriting a letter to a friend

3. making posters to present information

4. using a ruler, pencil, and graph paper to make a graph

5. making calculations using an instrument called the slide rule

Extending Your Experience

Answer the following in complete sentences. Use a separate sheet of paper.

1. Make a list of the ways you communicate with other people each day. Then, describe how technology has helped to make these activities better or even helped make them possible.
2. Think about some product of communication technology, such as the cell phone or personal computer. Describe how your life has been changed, for better or worse, by the development of that product.
3. Find out more about wireless communication and how it has revolutionized the communications industry.
4. Place yourself 100 years in the future. How might communication change with the discovery of new technologies?

Chapter 4 Review

Concept Review

A. Use these word parts and their meanings in Figure 4-29 to define the terms below.

Science Word Parts

- **meso-** of the middle
- **carn-** flesh, meat
- **chemo-** chemical
- **pyro-** fire
- **-zoic** of the animals
- **thermo-** heat
- **-nuclear** of the nucleus
- **-vorous** eating; feeding on
- **protero-** forward or ahead of
- **eco-** environment or habitat
- **paleo-** of the earliest
- **ceno-** newest; most recent
- **archeo-** of the ancient
- **-therapy** treatment
- **-lysis** loosening
- **-ology** the study of

▲ Figure 4-29

1. Cenozoic Era _____
2. pyrolysis _____
3. Mesozoic Era _____
4. thermonuclear _____
5. carnivorous _____
6. chemotherapy _____
7. archaeology _____
8. ecology _____
9. Paleozoic Era _____
10. Proterozoic Era _____

B. On a separate sheet of paper, draw an illustration of any house pet. Label the major parts.

C. Create a flowchart showing the order in which these steps occur in starting a car motor.

- The spark plugs fire and ignite gasoline.
- Put the key in the ignition.
- The starter motor turns engine.
- The engine runs.
- Turn the key.

D. List the sections of a formal scientific report. List them in the correct order.

11. _____
12. _____
13. _____
14. _____
15. _____
16. _____
17. _____

E. The table in Figure 4-30 shows the temperature increase of different colored cubes put under a heat lamp. Graph the data in Figure 4-31, putting numbers on the vertical axis and the color names on the horizontal axis. Give the graph a title.

Colored Cubes and Temperature Increases	
Color of Cube	Temperature Increase (°C)
White	6
Blue	10
Red	12
Black	20

▲ Figure 4-30 You can create graphs from information in a table.

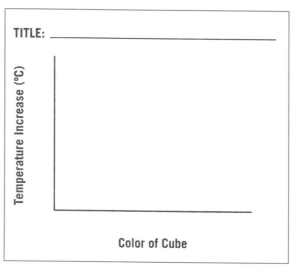

▲ Figure 4-31 Draw your graph.

Vocabulary Review

Complete the crossword puzzle using the Terms to Know from this Chapter.

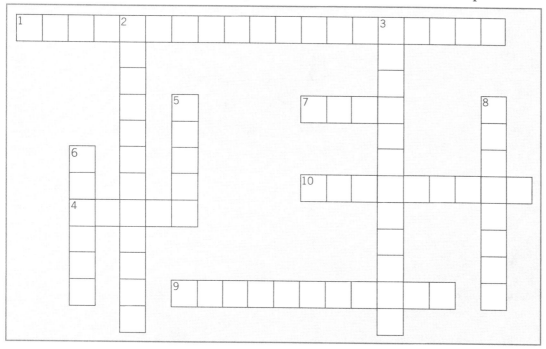

ACROSS

1. inside view of something cut through its long axis (2 words)
4. distance on a map representing a greater distance
7. two main lines of graph labeled with numbers or words
9. special terms in science
10. data represented by points that are connected (2 words)

DOWN

2. displays information as picture or diagram
3. inside view of something sliced across (2 words)
5. name of an experiment
6. an outcome reported to others
8. revolutionary means of communication

Chapter 4   Review   131

CAREERS ◆ in Science

There are thousands of careers related to the sciences. The list on this page gives the names of some of these careers. Use references to learn about them.

You may or may not have given much thought to what you want to do once you complete high school. Even though a career in the sciences may not be in your plans right now, one of these may interest you. Research the careers, paying close attention to the training and education required for each. Think of some good reasons why a particular career interests you.

Earth

aerospace technician
agronomist
air quality analyst
environmental officer
geologist
geophysicist
glaciologist
hydrologist
lapidary
metallurgist
meteorologist
mine inspector
mineralogist
oceanographer
paleontologist
petroleum geologist
seismologist
stonemason
volcanologist

EARTH ◆ This geologist is examining soil samples.

EARTH ◆ A weather report is being given by a meteorologist.

EARTH ◆ This hydrologist is taking water samples.

LIFE ✦ A botanist studies plants.

PHYSICAL ✦ An electrician is reading blueprints while at a work site.

Physical

automobile mechanic
biophysicist
chemical engineer
chemist
civil engineer
computer technician
draftsperson
electrician
electrical engineer
electronics technician
geochemist
heating/cooling technician
machinist
mechanical engineer
plumber
refrigeration technician
science teacher
welder
X-ray technician

Life

animal breeder
biologist
botanist
cardiologist
dental hygienist
ecologist
farmer
florist
genetic counselor
laboratory technician
nurse's aide
physical therapist
practical nurse
respiratory therapist
taxonomist
veterinarian
zoologist

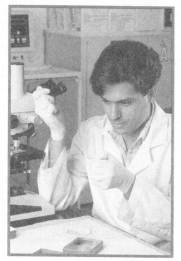
LIFE ✦ This biologist is analyzing DNA.

PHYSICAL ✦ A welder must practice safety when working.

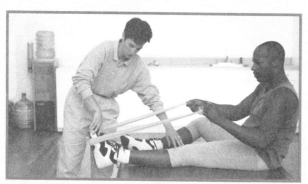
LIFE ✦ A physical therapist helps people recover from injuries.

PHYSICAL ✦ This chemist is conducting an experiment with four solutions.

Appendix Careers in Science | 133

MATHEMATICS SKILLS

Adding Integers

You can add integers with unlike signs on a number line.

Add. $^-5 + {}^+7$

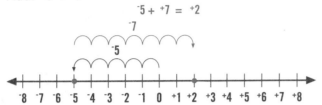

Subtracting Integers

To subtract an integer, add its opposite.

Subtract. $^-6 - {}^+2$

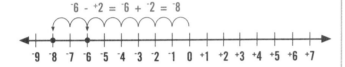

Multiplying Integers

When you multiply integers, you must decide if the answer is positive or negative.

If the signs of the integers are the same, the product is positive.

$${}^+5 \times {}^-4 = {}^+20$$
$${}^-5 \times {}^-4 = {}^+20$$

If the signs of the integers are different, the product is negative.

$${}^+5 \times {}^-4 = {}^-20$$
$${}^-5 \times {}^+4 = {}^-20$$

Dividing Integers

The rules for dividing integers are the same as the rules for multiplying integers.

If the signs of the integers are the same, the quotient is positive.

$${}^-36 \div {}^-9 = {}^+4$$
$${}^+36 \div {}^+9 = {}^+4$$

If the signs of the integers are different, the quotient is negative.

$${}^-36 \div {}^+9 = {}^-4$$
$${}^+36 \div {}^-9 = {}^-4$$

Adding Decimals

When adding or subtracting decimals, always be sure to line up the decimal points correctly.

Add 3.4 km, 20.95 km, and 153.6 km.

```
    3.4
   20.95
 +153.6
  177.85 km
```

Subtracting Decimals

Subtract 13.5 mL from 35.75 mL.

```
   35.75
 - 13.5
   22.25 mL
```

Multiplying Decimals

When multiplying or dividing decimals, it is not necessary to line up the decimal points.

Multiply 0.5 N by 11.25 m to find the amount of work done in joules.

$$W = F \times d$$
$$W = 0.5 \text{ N} \times 11.25 \text{ m}$$
$$W = 5.625 \text{ J}$$

Notice that the number of places to the right of the decimal point in the answer is equal to the sum of the places to the right of the decimal point in the numbers being multiplied.

Dividing Decimals

Divide 4.05 m by 0.5 m to find the mechanical advantage of a lever.

MA = effort arm length/resistance arm length
MA = 4.05 m/0.5 m
MA = 8.1

When dividing a decimal by another decimal, you must first change the divisor to a whole number. For example, change 0.5 to 5 by moving the decimal point one place to the right. You must also move the decimal point in 4.05 one place to the right. The result is 40.5 ÷ 5 = 8.1.

Changing a Decimal to a Percent

To change a decimal to a percent, multiply the decimal by 100%.

Find the efficiency of a machine if the work output is 5 J and the work input is 10 J.

efficiency = work output ÷ work input × 100%
efficiency = 5 J ÷ 10 J × 100%
efficiency = 0.5 × 100%
efficiency = 50%

Notice that when you multiply 0.5 by 100%, the decimal point moves two places to the right.

METRIC MEASUREMENTS

Length

The basic unit of length is the **meter** (m).

10 millimeters (mm) =	1 centimeter (cm)	
10 centimeters =	1 decimeter (dm)	= 100 millimeters
10 decimeters =	1 meter (m)	= 1,000 millimeters
10 dekameters (dkm) =	1 hectometer (hm)	= 100 meters
10 hectometers =	1 kilometer (km)	= 1,000 meters

Mass

The basic unit of mass is the **gram** (g).

10 milligrams (mg) =	1 centigram (cg)	
10 centigrams =	1 decigram (dg)	= 100 milligrams
10 decigrams =	1 gram (g)	= 1,000 milligrams
10 dekagrams (dkg) =	1 hectogram (hg)	= 100 grams
10 hectograms =	1 kilogram (kg)	= 1,000 grams

Volume

The basic unit of volume is the **liter** (L).

1 liter (L) =	0.001 cubic liter	
10 milliliters (mL) =	1 centiliter (cL)	
10 centiliters =	1 deciliter (dL)	= 100 milliliters
10 deciliters =	1 liter	= 1,000 milliliters
10 dekaliters (dkL) =	1 hectoliter (hL)	= 100 liters
10 hectoliters =	1 kiloliter (kL)	= 1,000 liters

GLOSSARY

This Glossary contains the important science words used in this book. Many of the words have a pronunciation guide in parentheses beside the word.

Many of the words in this glossary will be familiar to you. Some will not. If a word is part of your vocabulary, you can define it and use it in a sentence. Place a check (✓) beside each word in this glossary that is part of your vocabulary. As you come to know the meanings of other words in the glossary, check them off also.

A

apparatus (ap-uh-RAT-uhs) equipment, such as tools and devices, used to perform a task *p. 56*

axis a straight line used for reference *p. 116*

B

balanced equation an equation in which the number of atoms of each element on the left side of an equation is the same as the number of atoms of each element on the right side of the equation *p. 80*

bar graph a graph that uses bars of different lengths to show relationships; the bars can be horizontal or vertical *p. 116*

biased (BY-uhst) **sample** a sample that contains errors favoring one result over another *p. 60*

C

caption (KAP-shuhn) a short description under or beside a photograph, diagram, or illustration *p. 16*

cause an event or condition that brings about an action or result *p. 92*

census (SEHN-suhs) a study of each and every part of a group *p. 60*

chart a presentation of information in an easy-to-read format *p. 12*

chemical equation a shorthand way of describing how substances behave in a chemical reaction *p. 80*

chemical formula a shorthand way to write the name of a compound using chemical symbols *p. 80*

chemical symbol the shortened way of writing the name of an element *p. 80*

circle graph a circular diagram that shows relative sizes of the parts that make up the whole *p. 116*

classifications (klas-uh-fih-KAY-shuhns) groups or classes of objects with certain common characteristics *p. 76*

classify (KLAS-uh-fy) to arrange objects into groups based on similarities, or things in common *p. 76*

coefficient the number that shows how many molecules of a substance are involved in a chemical reaction *p. 80*

compare to tell how things are alike and different *p. 40*

comparison an examination of similarities and differences *p. 72*

compound a substance made up of two or more elements that are chemically combined *p. 80*

conclusion a decision reached about a question under consideration; a final answer or explanation *p. 96*

contour interval the difference in altitude between two adjacent lines on a topographic map *p. 112*

control a variable that does not change *p. 32*

control experiment an experiment in which a variable is held constant *p. 32*

convert to change from one unit to another *p. 48*

cross section an illustration of what an object would look like if part of it were cut away to show what is inside *p. 112*

crystal a solid with a definite shape *p. 84*

D

decimal a number written using the base ten *p. 48*

degree Celsius (SEHL-see-uhs) the basic unit of temperature *p. 40*

diagram a drawing that clearly shows how something is arranged *p. 16*

dictionary a book that gives pronunciations and definitions *p. 20*

different not alike *p. 72*

dominant gene a gene whose trait always shows itself *p. 24*

E

effect the action or result of a cause *p. 92*

element a substance that cannot be broken down into simpler substances through chemical change *p. 80*

encyclopedia a book that gives general and specific information on many topics *p. 20*

estimate (EHS-tuh-mayt) make an educated guess *p. 44*

estimate (EHS-tuh-mihtt) an educated guess *p. 44*

experiment an investigation or test *p. 20*

expert a person who knows a lot about a certain topic *p. 20*

extrapolation a prediction made for a value beyond observed data *p. 116*

F

flammable can catch fire easily and burn fast *p. 52*

G

general broad *p. 20*

generalization a broad, general statement about a topic that is true most of the time *p. 100*

gram the basic unit of mass *p. 40*

graph a visual way to communicate information *p. 116*

guide a source that can help direct a person's thinking *p. 84*

H

horizontal axis the crosswise axis of a graph *p. 116*

hypothesis (hy-PAHTH-uh-sihs) a possible answer to a scientific question based on information *p. 28*

I

illustration a drawing that communicates information, helping to make something clear *p. 112*

infer to form a conclusion based on evidence; to explain or interpret an observation *p. 24*

inference a reasonable conclusion based on information not directly observed *p. 96*

information a collection of facts and data obtained in any manner *p. 8*

Internet a network that allows computers around the world to communicate with each other *p. 126*

interpolation a prediction made for a value between observed data *p. 116*

interpretation (ihn-ter-pruh-TAY-shuhn) an explanation of what you have seen *p. 4*

K

key a table that will help you decode or interpret information *p. 84*

L

label an attached tag or highlighted section that identifies the product and provides information about the product and its use *pp. 8, 112*

length the distance from one point to another *p. 40*

line graph a graph that uses lines to show trends or changes over time *p. 116*

liter the basic unit of liquid volume *p. 40*

longitudinal section a cross section that is cut through the long axis of something *p. 112*

M

mass the amount of matter in something *p. 40*

measure to compare an unknown value with a known value using standard units *p. 40*

meter the basic unit of length or distance *p. 40*

O

observation (ahb-zuhr-VAY-shuhn) the practice of noting and recording facts and events *p. 4*

observe (uhb-ZERV) to pay attention to *p. 4*

P

pamphlet small, unbound booklet of printed material *p. 8*

pattern the repeated occurrence of some item *p. 88*

periodic repeated at regular intervals *p. 88*

periodic table a table of elements *p. 88*

prediction a statement made ahead of time about what you think might happen *p. 24*

Punnett square a chart that shows possible gene combinations *p. 24*

R

random sampling (RAN-duhm SAM-plihng) every part of the group has an equal chance of being included as part of the sample *p. 60*

recessive gene a gene whose trait is hidden when the dominant gene is present *p. 24*

record (REHK-uhrd) a lasting report that keeps information for later use *p. 64*

record (rih-KAWRD) to write down or save information in a permanent way *p. 64*

reference (REHF-uhr-uhns) something you know well that can be used to make a comparison *pp. 44, 72*

reference book any book that provides information *p. 20*

relationship a natural connection between two objects or events *p. 92*

report a detailed account of something *p. 122*

S

sample a study of just enough parts to understand the group *p. 60*

scale series of lines marked on a measuring tool; a distance used on a map to represent a greater distance on Earth's surface *pp. 40, 112*

similar the same *p. 72*

source person or place from which information is obtained *p. 8*

specific narrow *p. 20*

standard a value agreed upon by everyone *p. 72*

subgroup a small group within a larger group *p. 76*

subscript the number written to the lower right of a chemical symbol in a chemical formula *p. 80*

systematic sampling the use of a system to take the sample *p. 60*

T

technology the application of scientific knowledge to produce things needed or desired by people *p. 126*

temperature the measure of how hot an object is *p. 40*

terminology (as related to a science) the total of all the special words, or terms, that are used in that branch of science *p. 108*

transverse section a cross section at a right angle to a longitudinal section *p. 112*

U

unit of measure a standard quantity; a quantity measured by each unit agreed upon by scientists *p. 48*

V

valid generalization a generalization that can be supported as reasonable *p. 100*

variable something that may change, or vary *p. 32*

vertical axis the upright axis of a graph *p. 116*

volume the amount of space an object takes up *p. 40*

W

wired technology technology that sends information through wires *p. 126*

wireless technology technology that sends information from place to place without the use of wires *p. 126*

INDEX

A

Accuracy in record keeping, 64
Aerial photographs, 115
Apparatus. *See* Scientific apparatus
Ask-off competition, 23
Atoms, 81
Axis, 116
    horizontal, 116, 118
    vertical, 116, 118

B

Balanced equation, 80, 82
Balances, 58
Bar graph, 116
Biased sample, 60, 63
Buckminster Fuller, 73

C

Captions, 17
    defined, 16
    getting information from, 16–17
Card catalog, 15
Cartographer, 115
Cause, 92
Cause and effect, 92–95
Census, 60
Centimeter, 42, 75
Charts, 12–13
    recording findings in, 66
Chemical equations, 80, 81
    balancing, 82
Chemical formulas, 80, 81
Chemical shorthand, 80–83
Chemical symbols, 80
Circle diagram, 76
Circle graph, 116, 118–119
Classification, 76–79
Coefficients, 80, 81, 82
Communication
    graphs in, 116–121
    illustrations and models in, 112–115
    science vocabulary in, 108–111
    technology in, 126–128
    writing scientific reports in, 122–125
Comparisons, 31, 40, 44, 70, 72–75, 88
    making, to references, 73
    measurements in, 50
Compound, 80, 81
Computers, 126.
    See also Software for record keeping, 67

Conclusions, drawing, 96–99
    generalizations in, 100, 101
    in scientific reports, 122
Contour interval, 112, 113–114
Control, defined, 32
Control experiment, 32, 34
Control variable, 33
Conversions, 48
    in metric system, 49
Cross section, 112, 113
Crystals, 101
    identification of, 84, 85
Cubic meter (m^3), 40

D

Data, collecting, 38
Decimal system, 48, 51
Degree Celsius, 40
Diagrams, 17, 76
    circle, 76
    defined, 16
    getting information from, 16–17
    purpose of, 112
Dictionaries, 21, 23, 110
Differences, 72
Dominant gene, 24, 26

E

Earth, 77
Effect, 92. *See also* Cause and effect
Electrical equipment, rules for, 53
Electron microscope, 59
Element, 80
Encyclopedia, 21, 23
English System of Weights and Measures, 51
Equations, 80–82
    balanced, 80, 82
    chemical, 80, 81, 82
Estimates, 38
    graphs in making accurate, 121
    line graphs in, 118
    measurements and, 44–47
    in predictions, 44
    taking samples in, 61
    using references in making, 73
Experiments, 21, 64
    control, 33, 34
    defined, 33
    planning, 32–35
    reporting, 122–123
    in testing inferences, 24

variables in, 33, 34
Experts, 21, 23
Extrapolations, 116, 121
    as predictions, 121

F

Families, 89
Flammable materials, 52
Flowchart, 114
Formulas, 80–82
    chemical, 80
Fullerene, 73

G

Generalizations, 70
    in drawing conclusions or inferences, 101
    in forming hypotheses, 100
    making, 100–103
    valid, 102
General questions, 21
Genes
    dominant, 26
    recessive, 26
Genetics, 26
Geodesic domes, 73
Graduated cylinders, 41, 58
Gram, 40, 48
Graphs, 97, 106
    bar, 116
    circle, 116, 118–119
    line, 116, 118, 121
    in making accurate estimations and predictions, 121
    making and using, 116–121
    pie, 118
    steps in drawing, 120
Great Square, 116
Guides, 84–87
    in identification, 84–87

H

Horizontal axis, 116, 118
Hypothesis, 24
    defined, 28
    forming, for testing, 28–31
    generalizations in forming, 100
    illustrations to explain, 112
    in scientific reports, 122
    testing, 64

I

Identification,
    guides in, 84–87
Illustrations, 106, 112
    captions for, 17
    cross section, 113
    drawing, 112–115

longitudinal section, 113
transverse section, 113
Inferences, 24
    defined, 96
    generalizations in drawing, 101
    making, 24
Information
    defined, 8
    from diagrams, 16–17
    from line graphs, 118
    from measuring, 40–41
    organizing, 12–15
    from pictures, 12–13
    from sampling, 60–62
Insulators, 76
Internet, 10, 23, 126, 127
Interpolation, 116, 121
Interpretation of observation, 4–7
Introduction to scientific reports, 122
Investigations, 21, 70
    of cause-and-effect relationships, 95

L

Labels, 112
    defined, 8
    purpose of, 9
    terminology in, 110
Laboratory
    practicing safety in, 52–55
    setting up apparatus, 52
    working in, 56–59
Length, 40
Line graphs, 116, 118, 121
Liter, 40, 48
Longitudinal section, 112, 113

M

Maps, 114
    question, 20, 22
    topographic, 114
Mass, 40
Materials in scientific reports, 122
Measurements, 30, 38, 40–44
    estimating, 44–47
    information from, 40–41
    metric system in, 48–51
    referring to standards in, 75
Mendeleev, Dmitri, 88
Meter, 40, 48
Meter stick, 44
Metric ruler, 41, 42

Metric system, 48–51
  converting and comparing units in, 48–49
Millimeter, 42
Minerals, identification of, 84
Models, 112–115

N
Natural objects, identifying, 84
Newspapers, 15

O
Observations, 28
  illustrations in explaining, 112
  making, 4–7
  patterns in, 91
  in scientific reports, 122

P
Pamphlet, 8, 10
Patterns
  observation, 91
  recognizing, 70
  in science, 88–91
Periodic table, 88, 89, 102
Personal opinion, 31
Photographs, 115
  aerial, 115
  getting information from, 16–17
Pictographs, 121
Pictures, getting information from, 12–13
Pie graph, 118
Planning as scientist, 2–37
Predictions, 88
  defined, 25
  estimates in, 44
  extrapolations as, 121
  graphs in making accurate, 121
  making, 24–27
Prefixes, 108, 109, 111
Presentation software, 126
Problem solving, 2
Procedures in scientific reports, 122
Punnett square, 24, 26–27

Q
Question map, 20, 22
Questions
  answering, 28
  asking scientific, 20–23
  general, 21
  specific, 21

R
Random sampling, 60
Reading in science, 16–19
Recessive gene, 24, 26
Record keeping, 64–67
  accuracy in, 64
  computers in, 67
Records, 64–65
Reference book, 21
References, 72, 73, 75
  in making estimates, 73
  using, to estimate, 44–45
Relationships, 88, 92
  cause and effect, 92–95
Reports, 106
  scientific
    conclusions in, 122
    hypothesis in, 122
    illustrations in, 112–115
    introduction to, 122
    materials in, 122
    procedures in, 122
    results/observations in, 122
    safety role in, 122
    title of, 122
    writing, 122–125
Results in scientific reports, 122
Roots, 108, 111

S
Safety
  in laboratory, 52–55
  rules for, 52–53
  in scientific reports, 122
Safety goggles, 52
Samples, 60
  biased, 60, 63
  taking, in estimating, 61
Sampling, 60–63
  getting information through, 60–62
  random, 60
  systematic, 60, 61, 62, 63
Scale, 40, 41, 112, 113
Science
  asking questions in, 20–23
  drawing conclusions in, 96–99
  reading in, 16–19
  recognizing patterns in, 88–91
Science vocabulary, 108–111
Scientific apparatus, 56–59, 67
  inaccurate, 63
  safety in using, 54
  setting up, 52
Scientific discovery, 106

Scientific reports
  conclusions in, 122
  hypothesis in, 122
  illustrations in, 112–115
  introduction to, 122
  materials in, 122
  procedures in, 122
  results/observations in, 122
  safety role in, 122
  title of, 122
  writing, 122–125
Scientific skills
  asking questions, 20–23
  cause and effect, 92–95
  chemical shorthand, 80–83
  classifying, 76–79
  comparing, 72–75
  concluding, 96–99
  forming hypothesis, 28–31
  generalizing, 100–103
  graphs, 116–121
  guides, 84–87
  illustrations, 112–115
  information sources, 8–11
  laboratory work, 56–59
  measuring, 40–51
  models, 112–115
  observation, 4–7
  organizing information, 12–15
  pattern recognition, 88–91
  planning experiments, 32–35
  predicting, 24–27
  reading in science, 16–19
  recording, 64–67
  safety, 52–55
  sampling, 60–63
  scientific reports, 122–125
  technology, 126–128
  vocabulary, 108–111
Scientists
  communicating as, 106–131
  planning as, 2–37
  thinking as, 70–105
  working as, 38–69
Scrolls, 67
Seismographs, 39
Similarities, 72
Software. *See also* Computers
  presentation, 126
  spreadsheet, 126
  word-processing, 126
Solar system, 77
Sources for information, 8–11
Space probe, 59
Span, 45
Specific questions, 21

Spreadsheet software, 126
Standards, 72, 75
  measurements in, 75
Subgroups, 77
Subscript, 80, 81
Suffixes, 108, 109, 111
Symbols, 121
Systematic sampling, 60, 61, 62, 63

T
Tables, 97
Technology, 126–128
  wired, 127
  wireless, 127
Temperature, 40
Terminology, 108
  labels in, 110
Testing
  forming hypotheses for, 28–31
  reliable methods of, 30–31
Test tubes, 58
Thermometer, 41
Thinking as scientist, 70–105
Title of scientific reports, 122
Topographic maps, 114
  contour intervals on, 114
Transverse section, 112, 113
Triple-beam balance, 41

U
Units of measure, 48

V
Valid generalizations, 102
Variables
  control, 33
  defined, 32, 33
  planning, in experiment, 32–33
Vertical axis, 116, 118
Vocabulary, science, 108–111
Volcanologists, 39
Volume, 40

W
Weather satellites, 127
Wired technology, 126, 127
Wireless technology, 126, 127
Word parts, 108, 110, 111
Word-processing software, 126